SOFTWARE PROJECT SCHEDULING BY EVOLUTIONARY TECHNIQUES

DECEMBER 2018

TABLE OF CONTENTS

CHAPTER NO.	TITLE	PAGE NO.
	ABSTRACT	iii
	LIST OF TABLES	xii
	LIST OF FIGURES	xiv
	LIST OF SYMBOLS AND ABBREVIATIONS	xvi
1	INTRODUCTION	1
1.1	PROBLEM DEFINITION	2
1.2	OBJECTIVES OF THE RESEARCH	3
1.3	CONTRIBUTIONS OF THE RESEARCH	4
1.4	ORGANIZATIONS OF THE THESIS	4
2	OVERVIEW OF RESOURCE ALLOCATION AND TASK SCHEDULING IN SOFTWARE PROJECT DEVELOPMENT	7
2.1	SOFTWARE PROJECT	7
2.2	SOFTWARE PROJECT MANAGEMENT	8
2.2.1	Project Planning	9
2.2.2	Scope Management	10
2.2.3	Project Estimation	10

CHAPTER NO.		TITLE	PAGE NO.
2.3		SOFTWARE PROJECT SCHEDULING	13
2.4		PRINCIPLES OF SOFTWARE PROJECT SCHEDULING	14
2.5		FUNDAMENTALS OF SCHEDULING	14
	2.5.1	Classification	15
	2.5.2	Interdependence	15
	2.5.3	Time and Effort Allocation	16
	2.5.4	Validation Criteria	16
	2.5.5	Defined Responsibilities and Outputs	16
2.6		NEED FOR PROJECT SCHEDULING	16
2.7		RESOURCE ALLOCATION	17
2.8		RESOURCE MANAGEMENT	18
2.9		IMPORTANCE OF PROJECT SCHEDULING AND RESOURCE ALLOCATION IN SOFTWARE PROJECTS	18
2.10		SOFTWARE PROJECT SCHEDULING TECHNIQUES	20
	2.10.1	Program Evaluation and Review Technique (PERT)	22
	2.10.2	Critical Path Method (CPM)	23
	2.10.3	Resource-Constrained Project Scheduling Problem (RCPSP) Model	24
	2.10.4	Search-based Techniques	24

CHAPTER NO.		TITLE	PAGE NO.
	2.10.5	Event Based Scheduler (EBS)	25
2.11		SURVEY ON SOFTWARE PROJECT TASK SCHEDULING TECHNIQUES	25
2.11.1		General Methods for Software Project Scheduling	25
2.11.2		Resource Constrained Project Scheduling	35
2.11.3		Optimization Methods for Software Project Scheduling	39
2.12		SUMMARY	48
3		AN OPTIMIZED EVENT BASED SOFTWARE PROJECT SCHEDULING WITH UNCERTAINTY	50
3.1		INTRODUCTION	50
3.2		DESCRIPTION OF EMPLOYEES	51
	3.2.1	DESCRIPTION OF TASKS	53
3.3		PLANNING OBJECTIVE FUNCTION	54
3.4		EVENT-BASED SCHEDULER MODEL	55
3.5		ANT COLONY OPTIMIZATION ALGORITHM	56
	3.5.1	Construction of Task List	56
	3.5.2	Construction of Employee Allocation Matrix	57
	3.5.3	Pheromone Management	60

CHAPTER NO.		TITLE	PAGE NO.
	3.5.4	Local Mutation Process	60
3.6	RESULTS AND DISCUSSION		61
	3.6.1	Dataset Description	61
	3.6.2	Experimental Setup	61
	3.6.3	Project Completion Time versus Project Completion Cost	62
	3.6.4	Project Completion Time	64
	3.6.5	Project Completion Cost	65
	3.6.6	Project Allocation Time	67
	3.6.7	Productivity	68
3.7	SUMMARY		69
4	FUZZY C MEANS CLUSTERING WITH ACO FOR OPTIMIZED SCHEDULING AND RESOURCE ALLOCATION		71
4.1	Introduction		71
4.2	FUZZY C MEANS WITH ACO BASED TASK SCHEDULING AND RESOURCE ALLOCATION		72
4.3	RESULT AND DISCUSSION		78
	4.3.1	Project Completion Time versus Project Completion Cost	78
	4.3.2	Project Completion Time	80
	4.3.3	Project Completion Cost	81
	4.3.4	Project Allocation Time	83

CHAPTER NO.		TITLE	PAGE NO.
	4.3.5	Productivity	84
4.4	SUMMARY		85
5	FUZZY C MEANS CLUSTERING WITH ABC FOR FAST AND EFFICIENT SCHEDULING AND RESOURCE ALLOCATION		87
5.1	INTRODUCTION		87
5.2	FCM WITH ABC BASED TASK SCHEDULING		88
5.3	RESULT AND DISCUSSION		92
	5.3.1	Project Completion Time versus Project Completion Cost	92
	5.3.2	Project Completion Time	94
	5.3.3	Project Completion Cost	96
	5.3.4	Project Allocation Time	97
	5.3.5	Productivity	98
5.4	OVERALL RESULT AND DISCUSSION		100
5.4.1	Project Completion Time Vs. Project Completion Cost		100
5.4.2	Project Completion Time		102
5.4.3	PROJECT COMPLETION COST		103
5.4.5	Productivity		106
5.5	SUMMARY		107
6	CONCLUSION AND FUTURE WORK		109

CHAPTER NO.	TITLE	PAGE NO.
6.1	CONCLUSION	109
6.2	FUTURE WORK	111
	REFERENCES	**114**
	LIST OF PUBLICATIONS	**122**
	International Journals	**122**

LIST OF TABLES

TABLE NO.	TITLE	PAGE NO.
2.1	Comparison of Different Software Project Scheduling	43
3.1	Information of the Test Projects	62
3.2	Comparison of Project Completion Time versus Project Completion Cost	62
3.3	Comparison of Project Completion Time	64
3.4	Comparison of Project Completion cost	66
3.5	Comparison of Project Allocation Time	67
3.6	Comparison of Productivity	68
4.1	Comparison of Project Completion Time versus Project Completion Cost	78
4.2	Comparison of Project Completion Time	80
4.3	Comparison of Project Completion cost	81
4.4	Comparison of Project Allocation Time	83
4.5	Comparison of Productivity	84
5.1	Comparison of Project Completion Time versus Project Completion Cost	93
5.2	Comparison of Project Completion Time	95
5.3	Comparison of Project Completion cost	96
5.4	Comparison of Project Allocation Time	97
5.5	Comparison of Productivity	99

TABLE NO.	TITLE	PAGE NO.
5.6	Comparison of Project Completion Time versus Project Completion Cost	100
5.7	Comparison of Project Completion Time	102
5.8	Comparison of Project Completion cost	104
5.9	Comparison of Productivity	106

LIST OF FIGURES

FIGURE NO.	TITLE	PAGE NO.
2.1	Software Project Scheduling Process	21
3.1	Comparison of Project Completion Time versus Project Completion Cost	63
3.2	Comparison of Project Completion Time	65
3.3	Comparison of Project Completion cost	66
3.4	Comparison of Project Allocation Time	68
3.5	Comparison of Productivity	69
4.1	Flowchart of ACO Based task scheduling method	77
4.2	Comparison of Project Completion Time versus Project Completion Cost	79
4.3	Comparison of Project Completion Time	81
4.4	Comparison of Project Completion cost**Error! Bookmark not d**	
4.5	Comparison of Project Allocation Time	83
4.6	Comparison of Productivity	85
5.1	Comparison of Project Completion Time versus Project Completion Cost	94
5.2	Comparison of Project Completion Time	95
5.3	Comparison of Project Completion cost**Error! Bookmark not d**	
5.4	Comparison of Project Allocation Time	98
5.5	Comparison of Productivity	99
5.6	Comparison of Project Completion Time versus Project Completion Cost	101

FIGURE NO.	TITLE	PAGE NO.
5.7	Comparison of Project Completion Time	103
5.8	Comparison of Project Completion cost	**Error! Bookmark not d**
5.9	Comparison of Project Allocation Time	105
5.10	Comparison of Productivity	107

LIST OF SYMBOLS AND ABBREVIATIONS

ACO-L	-	ACO with local mutation procedure
ACO-UH	-	ACO with Uncertainty Handling
ACA	-	Ant Colony Algorithm
ACO	-	Ant Colony Optimization
ACS-SPSP	-	Ant Colony Optimization Software Project Scheduling Problem
ABC	-	Artificial Bee Colony
BPEP	-	Backward Peak Elimination Procedure
COCOMO	-	Constructive Cost Model
CCEA	-	Cooperative Co-Evolutionary Algorithms
CPM	-	Critical Path Method
DE	-	Differential Evolution
EBS	-	Event Based Scheduler
FP	-	Function Points
FCM	-	Fuzzy C means
KGA	-	Knowledge-Based Genetic Algorithm
LOC	-	Line of Code
MINSLK	-	Minimum Slack
MIP	-	Mixed Integer linear Programming
MOEA	-	Multi Objective Evolutionary Algorithm
PERT	-	Project Evaluation Review Technique
RCPSP	-	Resource-Constrained Project Scheduling Problem
SPMnets	-	software Project Management Net
TS	-	Tabu Search
TPG	-	Task Precedence Graph
TSRAP	-	Two Stage Resource Adjustment Procedure
WBS	-	Work Breakdown Structure

CHAPTER 1

INTRODUCTION

One of the issues the software development industry faces so far is the huge rate of failure of software enterprise. As per the October 2013 DDJ article on the Ambysoft declarations, IT Project Success review found that 58% of respondents esteemed being on schedule, 36% on budget, and 14% building to specification.

With a specific end intention to increase the success charge of software program initiatives, software managers must give importance to useful resource allocation and task scheduling. Both Resource Allocation and Task Scheduling play an essential role in software project development. Human Resource allocation involves the planning of all employees required for the project and assigning a developer to carry out a venture and trying to reply "Who will work on what?", while Human Resource scheduling includes specifying the time-frame, wherein a developer will work on a job and attempts to answer "Who will work whilst?".

The ambition of the Project Scheduling issue is minimization of project duration, minimization of project cost and maximization of product quality. The goal of resource allocation is to assign the personnel to appropriate tasks, so that the tasks can be performed efficiently.

Being different from projects in other fields, software tasks are humans-in depth sports and their related assets are especially human sources and require personnel with unique skills. Assigning personnel to the fine-outfitted responsibilities and human aid allocation has come to be a vital component in software program venture planning. Due to the importance and trouble in software task making plans, there's a growing necessity for developing effective laptop aided equipment in software venture planning in the recent years.

A project scheduling issue consists of defining which resources are used to perform which job and what is the duration of each. The allocation of developers to tasks and scheduling tasks are not two independent activities, but they are interrelated and needed to be worked on simultaneously.

1.1 PROBLEM DEFINITION

Many researchers take into account that practices related to making plan activities are the most important for the fulfilment of a software project and particularly due to the fact that these activities are required at the beginning of a project. Specifically, making plan activities require software project managers to adopt various budgeting and scheduling tasks so to determine how the software program product can be built, how plenty it'll value and the way lengthy it will take to supply it to the client. In order to reply these questions, two activities need to be done, namely resource allocation and task scheduling, during which software project managers decide what needs to be accomplished, whilst and by using whom. These two activities are considered extremely vital for the success of a software project because, on the one hand, erroneous task scheduling may cause great delays in delivery and budget overruns and, however, flawed aid allocation can cause an undesired low stage of exceptional in the software program products. The scheduling of the tasks present in the workload is a windy process. Many resource plans were affected by the unexpected joining and leaving event of human resources which may cause uncertainty. This uncertainty can also affect the quality of the project delivery. By appointing a developer simplest to the first allotted task until the completion of the same, may reduce the flexibility of human resources, even though the developer has the ability to do other based tasks. Optimized Event Based Scheduler handles this uncertainty and resource flexibility.

It is pretty commonplace that we need more time for scheduling if the developer record is enormous. Subsequently the search space is also big, and in the long run the resource allocation is not on time. To reduce the search space and to hurry up the retrieval of records, Fuzzy C Means Clustering with Ant Colony Optimization is implemented; Further, to growth the efficiency the Artificial Bee Colony algorithm is likewise applied.

1.2   OBJECTIVES OF THE RESEARCH

The optimal plan created in the beginning of project scheduling may have an expensive execution crumbling. The plan may be reproduced by simply reducing the impact of disruptions to the efficiency of the project. Planning process must create the new time tables that may vary as little as could be expected under the circumstances from the previous ones such that it ought to advance solidarity in elementary planning.

Certainly, the greatest asset for an enterprise is its human resources and it forms the foundation for all organization interactions. The resources must be utilized in a better way for completing the project based on an effective scheduling methodology by considering the following problems:

- Assign employees to appropriate tasks so that the undertakings must be possible effectively.
- Construct the productive schedule for the task's execution by considering the priority constraints so that the software construction cost can be reduced.

1.3 CONTRIBUTIONS OF THE RESEARCH

The contributions of the thesis can be listed as follows:

- The uncertainty handling mechanism is considered as an event in event-based scheduler. The uncertainty factors such as the unexpected absence of an employee and the termination of an employee are handled by proposing project scheduling and resource allocation approach which uses ACO and simulation tool in order to improve the project scheduling with uncertainty handling.
- The flexibility of human resources during scheduling is improved by proposing fuzzy clustering based ACO for software project scheduling. It considers the user acquaintance when the event-based scheduling task is performed.
- The overall search space is minimized by introducing the fuzzy clustering based ABC optimization algorithm for software project scheduling. Thus, the execution cost and time for software project scheduling are reduced by the proposed approach.

1.4 ORGANIZATIONS OF THE THESIS

The remainder of this work is structured as follows: To begin with, Chapter 2: Introduces about resource allocation and task scheduling in software development projects. It also presents the previous research of this thesis in the field of software plan scheduling calling different state-of-art algorithms such as general software project scheduling, resource constrained based software project scheduling and optimization based software project scheduling. Different advantages and shortcomings of each technique are also addressed.

Chapter 3: Describes the first contribution such as an optimized event based software project scheduling with uncertainty handling mechanism. It deals with the scheduling of tasks and allocation of resources under an uncertainty environment. The performances of the proposed technique are compared with the existing techniques such as ACO-L, KGA and TS.

Chapter 4: Presents the second contribution such as FCM clustering and experience based Ant Colony Optimization for software project scheduling with event based scheduler. It deals with human resources flexibility improvement during scheduling by using an ACO optimization algorithm which reduces the project allocation time, completion time and project completion expenditure. The performances of the suggested technique are correlated with the current technique.

Chapter 5: Explains the third contribution such as FCM clustering and experience based Artificial Bee Colony for software plan scheduling with event based scheduler. It deals with the reduction of search space during software project scheduling by using ABC optimization algorithm which reduces the project execution time, completion cost effectively. The performances of the proposed technique are compared with the existing technique. It also includes the overall Results and Discussions of the research work. It deals with the overall performances of the proposed techniques such as ACO-UH, FCM-ACO-UH and FCM-ABC-UH in terms of project completion cost, completion time, project allocation time and productivity.

Chapter 6: Concludes the research work with its findings and discusses the possible future enhancement on approaching the other variations of software project scheduling problems.

CHAPTER 2

OVERVIEW OF RESOURCE ALLOCATION AND TASK SCHEDULING IN SOFTWARE PROJECT DEVELOPMENT

2.1 SOFTWARE PROJECT

In software engineering, the job pattern is split into different parts such as software creation and software project management (Calderón, A., & Ruiz, M. 2015). A project is defined as a well-defined task and also a collection of various operations for achieving the targets like software delivery and software development. A software project may be exemplified as follows:

- Project is not day-to-day or routine operations.
- Each project has a distinct and unique goal.
- Project needs adequate resources in terms of material, time, finance, knowledge-bank and human power.
- It comes up with a start time and end time.
- It may end while its goal is achieved.
- Project may end while its goal is achieved. Hence, it is a temporary phase in the lifetime of an organization.

A software development is called as the complete procedure of software development from the process of requirement gathering to the process of testing and maintenance of software is carried out based on the execution methodologies, in a particular time period for achieving an intended software product.

2.2 SOFTWARE PROJECT MANAGEMENT

Software project management (Chemuturi, M., & Cagley, T. M. 2010, Cunha, J. A. O., et al 2016) is defined as the process of planning, staffing, monitoring, organizing, controlling and leading software projects. It is a sub-discipline of project management in which software projects are planned, implemented, monitored and controlled. The effectiveness of project management is the most essential for assuring the success of any substantial activity. The required factors for project management are knowledge, skills, goals and personalities. In software organization, it is the most significant part for delivering the quality product, keeping the cost within the client's budget constraint and delivering the project as per the schedule. Therefore, software project management is an essential for incorporating user requirements which includes price and time constraints.

A software project manager is a person who undertakes the responsibility of executing the software project. They are the most responsible persons for scheduling and planning of software project development. In addition to that, they lead the software development team, and they are the interface with senior management, initiator and suppliers. The software project manager's job makes sure that the software project delivering and project constraints in time. Software project management is a technique of arranging all activities related to the projects and its parts. Based on the establishment of project management, it consists of five phases are proposal writing, project planning, project scheduling, project tracking, personnel selection and evaluation and project report writing. Project management process can be applied for all project types but it is mostly utilized to handle the complex processes of software development projects. It is an application of techniques, knowledge, tools and skills to project activities for obtaining the project requirements. The software project management is required since professional software engineering is always subject to organizational budget and schedule constraints. The major activities in software project management are as follows:

- Initiating
- Planning
- Executing
- Monitoring and Controlling
- Closing

2.2.1 Project Planning

Software project planning (Greer, D., & Conradi, R. 2009, NadanaSundaram, P. V., & Iyakutti, K. 2015) is the task which is performed before the production of a software project actually starts. It is also defined as the discipline for stating how to complete the project within the particular time frame along with defined stages, and designated resources. It consists of various activities which are given as follows:

- Initiating objectives
- Determining deliverable
- Planning the schedule
- Creating supporting plans

The most preferable tools for project planning are Gantt chart and PERT chart.

2.2.2 Scope Management

It defines the scope of the project which includes all the activities, processes required to be achieved for making the deliverable software product. Scope management (Khan, A. 2006) is essential since it generates the boundaries of the project by clearly defining what would be prepared in the project and what would not be completed. This makes up for the project for containing the limited and quantifiable tasks which may easily be documented and in turn removes the cost and run time. During software project management, the following activities are essential.

- Defining the scope
- Deciding its verification and control
- Dividing the project into different smaller parts for ease of management
- Verifying the scope
- Controlling the scope by incorporating the modifications to the scope

2.2.3 Project Estimation

An accurate estimation of different measures is necessary for an effective management (Nasir, M. 2006). With accurate estimation, the manager can manage and control the project more effectively and efficiently. Project estimation may involve the following:

- Software size estimation (Alves, L. M., et al 2014): Software size may be estimated by finding the number of function points in the software which may vary according to the software or user requirements. On other hand, it can be estimated in terms of Kilo Line of Code (KLOC) that depends on the coding practices.

- Effort estimation (Ono, K., et al 2016, Sehra, S. K. et al 2014): The software managers in a software project can estimate the efforts in terms of employees for producing the software. The software size must be known for the estimation of effort. The effort can be obtained by the organization's historical data software size or manager's experience that is transformed into by using some standard formulae.
- Time estimation (Manapian, A., & Prompoon, N. 2014, Štolfa, J., et al 2012): After the estimation of efforts and size of software the time needed for producing software is estimated. Efforts required are separated into sub-categories as per the interdependency of different software components and requirement specifications. By using Work Breakdown Structure (WBS) software tasks are separated into smaller events, activities or tasks. Then the separated tasks are scheduled in calendar months or on a day-to-day basis. The summation of time needed to complete all the tasks of software project in hours or days is the total time spent for completing the project.
- Cost estimation (Li, Y. F., et al 2008, Klakegg, O. J., & Lichtenberg, S. 2016): The estimation of cost is the most difficult process because of it depends on many elements. In order to estimate the cost of a project, the following elements should be considered:
    → Travel involved
    → Training and support
    → Additional software or tools, etc
    → Size of software
    → Hardware
    → . Communication
    → Skilled employees with task-specific skills
    → Software quality

Empirical estimation technique and decomposition technique are used to estimate cost, effort, time and size of software.

The empirical estimation technique making estimation by using empirically derived formulae. The empirically derived formulae are based on FP or LOC.

- Putnam model (Suelmann, H. 2014): This model is extended based on Norden's frequency distribution such as Rayleigh curve. Putnam model maps efforts and time needed with size of software.
- COCOMO (Islam, S., et al 2016): COCOMO is abbreviated as Constructive Cost Model. It classified the software product into three categories, like embedded, organic and semi-detached.

Another technique is decomposition technique which considers the software as a product of different compositions. This technique involves two major models which are described as follows:

- Function Points (FP): An estimation is achieved on behalf of number of function points in the software product.
- Line of Code (LOC): The estimation is achieved on behalf of number of line of codes in the software product.

One of the most demanding tasks for software managers is software project scheduling. It is an activity that distributes estimated effort across the planned project duration by allocating the effort to specific engineering tasks. On the other hand, project scheduling involves separating the total work involved in a project into separating activities and judging the time required for completing these activities. Managers should also estimate the resources which are required for completing each task.

2.3 SOFTWARE PROJECT SCHEDULING

Software project scheduling (Peixoto, D. C. et al 2014) is referred to as an activity that distributes the estimated effort across the planned project duration by allocating the effort to specific software engineering tasks. The main intend of a software project scheduling is developing the set of engineering task which will allow completing the job in time. The responsibilities for each task in software development are assigned by developing a network of software engineering tasks. In addition to that, by developing a network of software engineering tasks execution can be controlled and tracked and also, if necessary, risks can be adapted.

Usually, developing large software systems involve a different and large number of interdependent tasks. Without scheduling such tasks, it is more complex to manage and understand the tasks. Moreover, without scheduling software project may not be evaluated in practice. The steps for performing the project scheduling after effort and size estimation, includes allocation of effort and duration to each task and design of a task network for enabling the team to meet the established delivery deadline. Software project scheduling consists of several advantages such as enhancing visibility, saving time, uncovering problems, building consistency, fixing problem, etc. Normally, the project scheduling is signified as a set of charts showing the staff allocation, work breakdown structure and activities dependencies. For scheduling the project, the following processes are necessary.

- Breaking down the project tasks into smaller and manageable form.
- Finding out various tasks and correlating them.
- Estimating the time frame required for each task.
- Dividing time into work-units.
- Assigning adequate number of work-units for each task.

- Determining the total time required for the project from beginning to end.

2.4 PRINCIPLES OF SOFTWARE PROJECT SCHEDULING

The principles of software project scheduling are listed as below.

- Compartmentalization: The project should be disintegrated into manageable tasks and activities.
- Time allocation: Each task should be allocated a number of time units, also possibly a start date and a completion date.
- Interdependence: The relationships between the tasks have to be established because of some activities will depend on other, when other activities may occur independently.
- Effort validation: Each project has a described number of staff. The project manager should ensure that no more than the allocated number of people have been scheduled at any given time.
- Defined responsibilities: Each task that is scheduled must be given to a specific team member.
- Defined outcomes: Each task that is scheduled must have a defined result.
- Defined milestones: Each task or group of tasks must be associated with the milestone. A milestone is accomplished when one or more work products have been reviewed for quality and approved.

2.5 FUNDAMENTALS OF SCHEDULING

The main objective of software project schedule is to determine the duration of the software project and the phases within the project. A software project schedule enables the distribution of estimated efforts to be spent in performing the critical activities. The basic principles for determining how the software project schedule is created are given below.

- Classification
- Interdependence
- Time and Effort allocation
- Validation criteria
- Defined responsibilities and outputs

2.5.1 Classification

Similar tasks are grouped together during the management of software project to complete a project successfully. Work Breakdown Structure (WBS) is widely used tool to group the similar tasks and to decompose the techniques. By utilizing these tools, a software project can be split into different phases, those different phases can also sub divided into activities. Based on the arrangement of the phases, the software project schedule is prepared.

2.5.2 Interdependence

A software project is a collection of multiple phases and each phase of the software project is a collection of multiple activities. Each activity of the software project is connected with each other but is treated separately. The sequence and interdependence of activities can be easily identified. Some activities get input from other activities to complete the software development while the some activities may be completed without any input from other activities.

2.5.3 Time and Effort Allocation

Each activity in a software project needs a particular amount of effort and time to complete activities. It can be managed by assigning start date and end date of each activity. Most software projects operate with time and effort constraints. Hence, managing the available resources is very essential for a software project manager.

2.5.4 Validation Criteria

The validation criteria for time and effort allocation in a software project are identified. The optimal level of resources is guaranteed by allowing the determination of the validation criteria which is available for the particular activity. Resources allocated are usually more than the actual requirement.

2.5.5 Defined Responsibilities and Outputs

The responsibilities of all people assigned to a software project define the hierarchy in the development team. The results are also defined for each activity. This helps in identifying the results expected at the end of each activity. Each role is linked to the outputs. When each role is linked to the expected outputs, each person's effort and the progress of each activity towards closure can be tracked.

2.6 NEED FOR PROJECT SCHEDULING

Software projects have a tendency for escaping of control since multiple activities need to be monitored, tracked, and controlled. While the project is out of control, the original deadlines, the budget, and the effort required overshoot the initial estimates. This does not have a great impact on the product, but also the credibility of the development team. The following are the requirements for developing a project schedule.

- Project scope
- Sequence of activities
- Tasks should be grouped into conception, definition and planning, launch, performance, and close
- Task dependencies map
- Critical path
- Project milestones

2.7 RESOURCE ALLOCATION

Resource allocation is the process of assigning and scheduling the available resources in the most effective and economical way, Projects will always require resources and resources are usually scarce. The task therefore lies with the project manager for determining the appropriate timing of those resources within the project schedule. Resource allocation in project management is an ongoing process that should be efficiently managed in alignment with available resources as well as the organizational strategy.

In other words, resource allocation is the process and strategy involving a company deciding where scarce resources should be utilized in the production of goods or services. A resource can be considered as any factor of production, which is utilized for producing goods or services. In project management, resource allocation is the scheduling of activities and the resources required by those activities while taking into consideration both the resource availability and the project time. In an organization, resource management is the efficient and effective development of an organization's resources when they are required. Such resources may include the financial resources, inventory, human skills, production resources or information technology.

2.8 RESOURCE MANAGEMENT

The elements which are utilized in the software development process are called as resource for project (Hartmann, S., & Briskorn, D. 2010, Anwar, Z., et al 2014). The resource includes the software libraries, human resource and productive tools. Resources are available in limited in quantity and stay in organizations as pool of assets. The lack of resources hinders the product development and it leads to lag behind schedule. Due to lack of resources, additional resources will be allocated this will increase the cost of development in the end. Thus, it is more important to allocate and estimate adequate resources for the project. Resource management consists the follows:

- Describing proper organization project through developing a project team and allocating the responsibilities to each team member involved in the project.
- Finding resources needed at a specific stage and their availability.
- Managing the resources by generating resource request while they are required and reallocating them when they are not required.

2.9 IMPORTANCE OF PROJECT SCHEDULING AND RESOURCE ALLOCATION IN SOFTWARE PROJECTS

While a project plan contains more than just a schedule, the schedule is arguably the most important aspect of that plan. For the project team, the schedule defines what they need to do and when they need to do it. Outside the project team, it is the most visible element of the plan, and many consider it to be the plan. The importance of project scheduling is as follows:

- For constructing the complex system in such a way that much software engineering tasks occur simultaneously.
- The outcome of work performed during a single task may have the profound effect on the work to be conducted in another task.
- These interdependencies are very difficult to comprehend without the scheduling process.
- It's also virtually impossible for assessing the progress of the software project without the detailed scheduling, be it moderate or large.

For organizing and completing the projects in a timely, qualitative and financially responsible manner, we need to schedule the projects carefully. Effective project scheduling plays a crucial role in ensuring the project success. To keep projects on track, set realistic time frames, assign the resources appropriately and manage quality for decreasing the product errors. This typically reduces costs and increases customer satisfaction.

Project scheduling ensures one task gets completed in a qualitative manner before the next task in the process begins. By assuring that quality measures meet the expectations at every step of the way, we ensure that managers and team members address problems as they arise and don't wait till the end. No major issues must appear during the completion since we have established quality controls from the very beginning of the scheduling process. Effective project managers understand that ensuring quality control involves managing the risks and exploiting the opportunities for speeding up the schedule, when possible to beat the competition and achieve or maintain the competitive edge with a more reliable product.

Scheduling is the process which actually manage the projects. Without scheduling, nothing or no one can manage the project and hence leads to failure of the project. Scheduling describes the guidance and pathway for a project to run. It defines specific milestones and deliverable which need to be achieved on a timely basis for the successful completion of the project. Monitoring the schedule provides an idea of the impact the current problems have on the project, and provides the opportunities for enhancing or reducing the scope of the milestone or phase in the project.

It also provides the medium for continuous feedback on how the project is progressing and if there are problems that require to be dealt with or if the client requires to be informed about the delay in delivery. Project Scheduling affects overall finances of the project. Time constraints make the project managers to schedule resources effectively. This is absolutely true because the resources should have highly knowledge and specialized skills in order to complete the task or costly materials are needed. It is notable that completing a software project within a short time frame consumes more cost because of expedited materials or additional resources are required. By scheduling project accurately, projecting accurately and estimate realistically that prevent last-minute orders which drive up costs.

2.10 SOFTWARE PROJECT SCHEDULING TECHNIQUES

Different software project scheduling techniques (Kaur, R., et al 2013, Suri, B., & Jajoria, P. 2013) are given below.

- Work Breakdown Structure (Brotherton, S. A. 2008): There are different ways of breaking down the activities in a project, but the most usual is into work packages, tasks, deliverable and milestones.
- Activity Charts: An activity network is a labeled graph along with the nodes corresponding to activities, arcs labeled with estimated times, and activities are connected as there is a dependency among them.
- Project Evaluation Review Technique (PERT) (Xiang, C., & Yue, W. 2009)

- Gantt chart: Gantt charts are a type of bar chart in which time is plotted on x-axis and bars on y-axis for each activity.
- Critical Path Method (CPM) (Ren, Y., et al 2010): The critical path is the sequence of activities that takes the longest time for completion.

The following figure 2.1 shows the general block diagram of software project scheduling process.

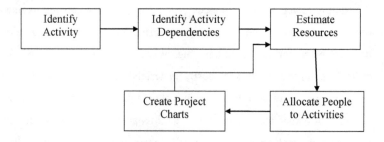

Figure 2.1   Software Project Scheduling Process

- Identify Activity: Identifying the specific activities is a process which must be performed to produce the various project deliverable.
- Identify Activity Dependencies: Documenting and identifying interactivity dependencies.
- Allocate Resources: Allocate the resources and calculating the number of periods which will be needed to complete an individual activity.
- Create Project Charts: In order to analyze the resource requirements, activity sequences and activity duration by creating project charts for creating the project schedule.
- Allocate People to Activities: The people are allocated to activities based on various activities.

There are different techniques developed for software project scheduling in software engineering. Among those techniques, the most notable project scheduling techniques are Event-based scheduler software Project Management Net (SPMnets) model, Critical Path Method (CPM), Search based techniques, Resource-Constrained Project Scheduling Problem (RCPSP) model, Program Evaluation and Review Technique (PERT) model. The main objective of these techniques is delivering the project on time, effort distribution and managing human and technical difficulties. Each technique is designed on the particular principle for attaining the end user requirements and managing the projects.

2.10.1 Program Evaluation and Review Technique (PERT)

PERT (Yunning, Z., & Xixi, S. 2010) is a technique to analyze the tasks involved in the software development process, specifically the time required for completing each task and for identifying the minimum time required for completing the entire project.

PERT chart explicitly describes and makes visible dependencies between the work breakdown structure elements. It helps identification of the critical path and makes this visible and additionally helps identification of late start, early start and lack for each activity. PERT provides for probably decreased project duration because of higher understanding of dependencies main to advanced overlapping of activities and duties where viable. The large quantity of project data can be prepared for utilizing it in decision making.

The major drawbacks of PERT are such that it is lack of the timeframe on most PERT, not easily scalable for smaller projects, CPM charts makes it difficult to show status although colour can help. Network charts be likely to be unwieldy and large requiring different pages for printing and requiring special size paper. At the same time as the PERT charts end up unwieldy, they may be not used for managing the project and there may be doubtlessly loads or hundreds of activities and individual dependency relationships. During project execution, a real project can never be executed if it was planned uncertainly. It is able to be ambiguity on account of subjective estimates which are prone to human mistakes or it is able to be variability springing up from sudden activities or risks. The most important motive that PERT can offer inaccurate information about the project completion time which is due to this time table uncertainty. This inaccuracy is huge sufficient for render such estimates as no longer beneficial.

2.10.2 Critical Path Method (CPM)

A CPM (Stelth, P., & Roy, G. L. 2009, Ren, Y., et al 2010) is a project management tool which is used for formulating the time frame for a project in order to determine where potential delays are most likely to occur. The process includes the step-by-step process which provides the developer with the visual representation of potential bottlenecks throughout the course of the project. CPM is useful in monitoring and controlling projects and it is widely used in industry since it determines slack and floats times. A project manager may determine the actual dates for each activity and compare what must be happening to what is taking place and react correspondingly. CPM is used to determine the project duration, which minimized the sum of direct and indirect costs. It is used for evaluating which activities may run parallel to each other and may define the multiple and equally critical paths. The activities and their outcomes may be considered as a network.

The major limitations of CPM are such that CPM's are complicated and the complexity is increased in larger projects. It does not handle the scheduling of personnel or the allocation of resources. The critical path is not always accurate and it requires being determined carefully. Moreover, the estimation of activity completion time has high complexity.

2.10.3 Resource-Constrained Project Scheduling Problem (RCPSP) Model

Resource-constrained project scheduling method (Chand, S. 2016, Madan, M., & Madan, M. R. 2013) involves the project activities scheduling which is subjected to precedence and resource constraints for meeting the objectives in the possible way. The RCPSP involves the scheduling of a project in which its duration is subjected to zero-lag finish-start precedence constraints of the PERT or CPM type and constant availability constraints on the required group of renewable resources.

2.10.4 Search-based Techniques

Search based techniques are used for optimizing the software project resource allocation (Bibi, N., et al 2014). The search based techniques can be applied for optimizing resource allocation in a software engineering project. The three notable search based techniques are encoding method, the hill climbing approach and the simulated annealing approach. Outcomes of search based techniques may change the size of the project teams. When for a small overall staffing level, double-sized teams does not improve the performance, and for a large overall staffing level, double sized teams are effective.

2.10.5 Event Based Scheduler (EBS)

Event Based Scheduler (EBS) (Vairagade, R. S., et al 2016) is used for developing the flexible and effective model for software project planning. This approach represents the plan by a task list and the planned employee allocation matrix. In this method, both the problems of employee allocation and task scheduling are considered. The starting time of the software project such as the time when the employee join or leave and the time when resources are released from the finished tasks are regarded as events.

The primary concept of EBS is regulating the allocation of employees at events and keeping the allocation unchanged at non-events. With this strategy, the EBS method is utilized for enabling the modelling of resource conflict and task pre-emption and preserving the flexibility in human resource allocation of software project.

2.11 SURVEY ON SOFTWARE PROJECT TASK SCHEDULING TECHNIQUES

This section provides the details about the previous researches which are related to software project task scheduling based on different methods and different techniques. Such different types of techniques for software project task scheduling problem are discussed in briefly below.

2.11.1 General Methods for Software Project Scheduling

A novel Multi Objective Evolutionary Algorithm (MOEA) based dynamic scheduling method was proposed (Shen, X. et al 2016) for dynamic project scheduling. It generated a robust schedule predicatively and adapted to conventional scheduling algorithms in response to critical dynamic events during project execution. The first contribution of this MOEA based dynamic scheduling method was to model the software project scheduling problem in a dynamic and uncertain environment with the consideration of multiple objectives and constraints. The second contribution was a mathematical model for MODPSP was constructed and the third contribution MODPSP was solved by designing MOEA based dynamic scheduling method. But the simplifications of this method were not valid.

A tool named IntelliSPM was proposed (Stylianou, C. et al 2012) to assist software project managers in completing the most complex activities of project scheduling and staffing. This tool comprised of various optimization mechanisms. It supports schedule based optimal staffing using a multi objective Genetic Algorithm approach. Moreover, along with the staff based optimal project scheduling a hybrid optimization approach was employed through a multi objective Genetic Algorithm approach followed by a single objective Particle Swarm Optimization or Genetic Algorithm. The main purpose of these processes was to allow the project managers to create better staffing strategies in terms of personality traits and developer skill levels, along with that of the optimal project schedules within the shortest make span. But this tool offered limited ability for project managers to control the objectives.

A knowledge based evolutionary approach (Yannibelli, V., and Amandi, A. 2011) for software development project scheduling process. This approach assists the project managers at the early stage to schedule the project at an early stage. The proposed algorithm designed feasible schedules for the project and computed each designed schedule according to the different optimization objective. This approach was developed based on available knowledge about the employees in each schedule. This algorithm effectively computed the set of employee in each schedule of the assigned project activities. It was computed based on the effective level of each employee in the set of the employee. Finally, the effective level of an employee was defined based on the available knowledge raised from historical information about the employee participation in already executed projects.

A time line based model for software project scheduling was proposed (Chang, C. K. et al 2008) with the genetic algorithm. This model was capable of more realistically simulating real world situations. It was described along with a new Genetic Algorithm which produced optimal or near optimal schedules. The internal genome representation in Genetic Algorithm which means the model parameters and structures were not improved in the Genetic Algorithm. Complex Project management schemes were not considered in the old model of Genetic Algorithm. So, in the time line based model more parameters were added to enhance the user's control. In addition to that, software project cost estimation techniques were added into the model.

A reinforcement learning was presented (Padberg, F., and Weiss, D. 2011) to compute the optimal scheduling policies for software project scheduling problem. This approach was based on a stochastic, formal scheduling model which captured the strong feedback between tasks in software development. The optimal policy was computed for the sample projects and simulated the project and analyzed the task assignments which were made by the optimal policy. The optimal policies were found and assigned tasks based on the characteristics of software design and the past performance of the developers. In addition to that, the problem of scheduling large or strongly coupled components in the tasks was addressed in this paper. The approaches to the optimization of the large projects were sketched.

A novel approach with an Event based Scheduler (EBS) was developed (Chen, W. N., and Zhang, J. 2013) for software project scheduling and staffing. The proposed approach signified a planned employee allocation matrix and a plan by a task list. Through this approach the issues of employee allocation and task scheduling were considered. In EBS, events are considered based on the starting time and ending time of the project. The main intention of EBS was to adjust the allocation of non employees at nonevents and allocation of employees at events. Based on this strategy, the proposed EBS approach enabled the modeling of task preemption and resource conflicts and defended the flexibility in human resource allocation. By using Ant Colony Optimization (ACO) the planning problem in the software projects were solved. The proposed approach yielded better plans with more stable workload assignments and lower cost than the conventional approaches. The human resources were not effectively managed in this approach which plays an important role in the software project management.

Improved Evolutionary algorithm was proposed (Minku, L. L. et al 2014) for project scheduling problem. This proposed approach overcame issues related to the hit rate and to improve the solution quality of the software projects. Improved Evolutionary algorithm comprised of two main points. In the first point the problem of overwork was addressed and the employee dedications were normalized which offered an alternative solution to the repair mechanism that tried to minimize the dedication values across the board. In the other point, a new type of penalty was introduced in the computation of completion cost and time that offered a clear gradient for searching the feasibility. This proposed algorithm, a single objective is considered for software project scheduling problem. But the multi objective will provide better solution for software project scheduling.

An approach using Bayesian approach was introduced (Khodakarami, V. et al 2007) to address the uncertainty and causality problem in project scheduling. The proposed approach was adapted as one of the best scheduling algorithms called as Critical Path Method (CPM) and combined this method into the Bayesian network which is called as explicit uncertain model. Bayesian networks were widely utilized in a wide range of decision support applications, but its application to the project management is novel. This approach empowered the conventional CPM to tackle the uncertainty problem and also provided explanatory analysis to handle, represent and elicit the different sources of uncertainty in project planning.

Several Meta heuristic algorithms were proposed (Chicano, F. et al 2011) to address the problem of software project scheduling. The Meta heuristic algorithms approached the software project scheduling problem with their multi objective formulation, where both the project cost duration and project cost had been minimized. In this approach, five multi-objective Meta heuristics were used called as NSGA-II, SPEA2, PAES, MOCell, and GDE3. These Meta heuristics covered different project scenarios with different levels of difficulty. From the analysis, it is known that the PAES, MOCell, and GDE3 are utilized on the software project scheduling problem and the NSGA-II and SPEA2 were widely utilized in multi objective algorithms. The analysis of solutions of different algorithms determined the correlation between the features and regions in the objective space where these solutions were located.

A procedure for intelligent software project scheduling and with team staffing was proposed (Stylianou, C., and Andreou, A. 2011) with genetic algorithms. The proposed approach solved the software project scheduling problem and team staffing through adopting genetic algorithm as an optimization technique. By using the genetic algorithm, a project's optimal schedule was constructed and it assigned the most experienced employees to the tasks. The corresponding objective functions were utilized in the genetic algorithm to find the optimal solutions for projects of varying sizes. By combining different objective functions, genetic algorithm faced difficulties to find the optimal solutions. From the analysis it was found that the genetic algorithm couldn't minimize the idle gaps or it cannot generate a conflict free schedule. Moreover, there will be a local optima and convergence problem in the genetic algorithm which affects the performance of determination of optimal solution of software project scheduling.

Differential Evolution (DE) algorithm was presented (Amiri, M., and Barbin, J. P. 2015) to optimize the software project scheduling problem. It is one of the meta-heuristic algorithms which are a population based probabilistic search algorithm that solves optimization problems. This algorithm utilized the distance and direction information from the current population which carries out the search operations. It searched the process in the direction of coordinate axes of optimistic variables and also changed the direction of the coordinate axes in the right direction. It started the search process from the random initial population and there three operators are processed to optimize the software project scheduling problem. The processes are mutation, selection and integration. In addition to that, three control parameters number of population, possibility of integration and sole factors decided the optimization of software project scheduling problem.

An algorithm called as Cooperative Co-Evolutionary Algorithms (CCEA) was proposed (Ren, J. et al 2011) to search based the software project management that optimized both the work package and the developer's team staffing scheduling. The co evolutionary algorithm evolved two populations. One of the two populations is signified by the work package arranged in a queue that allotted their assignments to teams. Another population of two populations signified the developer's distribution among teams. The proposed CCEA project planning and staffing was compared with the random search and single population optimization using genetic algorithms. From the analysis, it was found that the CCEA outperforms the single objective optimization using genetic algorithm and random search in terms of the best solutions and was proposed in terms of project completion time and a smaller number of evaluations. But the proposed algorithm has the problem of communication overhead and schedule robustness.

A method was presented (Mathew, J. et al 2016) for repetitive project scheduling with the objective of minimizing project cost, project duration and both. In this paper, a penalty cost was added to the total cost when a particular activity of the project was not completed within the due date of the activity of that project. It considered the constraints of precedence relationships among the project units and constraints of precedence among the activities and constraints of the due date in which work should completed for each activity in every unit. It helped the project manager for selecting the best crew options which optimized the project cost and project duration in repetitive project works. Thus, this proposed method optimized the project duration and project cost with better solutions.

A mean variance and a mean semi variance models were proposed (Huang, X. et al 2016) to solve the joint problem of optimal project selection and scheduling. In this paper, the joint problem was discussed in the situation where the net cash and initial outlays of projects were fed by the expert's estimates due to lack of historical data about the project and the employees involved in the project. These parameters were defined by using uncertain variables and the utility was justified in this paper. The proposed model considered the relationship and time sequence between the projects. The complex problems were solved by the method introduced which computes the uncertain lower partial semi variance and higher partial semi variance values. In addition to that, a hybrid algorithm was introduced that combined the genetic algorithm with cellular automation. But this model failed to take more life setting into account to solve the optimal project selection and scheduling problem.

A new bi-level model with multiple decision makers was proposed (Xu, J. et al 2015) for software project scheduling problem. This model solved the project scheduling problem by considering the project owner interest along with the project contractor interest. In this model, the project contractor was treated as the follower and the project owner was treated as the leader. The project owner has two objectives are to maximize the profit with time and to minimize the make span with investment and time constraints but the project contractor has only one objective, is that to minimize the cost with precedence and cost constraints. The proposed fuzzy random simulation based bi-level global local neighbour particle swarm optimization technique was used to solve the multiple decision maker projects scheduling problem. The PSO has some convergence problem which is the major drawback of the bi-level model.

A Mixed Integer linear Programming (MIP) based heuristics was proposed (Zimmermann, A. 2016) with work content constraints for project scheduling problem. Initially the project activities were arranged in a precedence feasible list based on the priority rules. Then, those listed activities were scheduled interactively as they were arranged in a list. Thus, the sequential scheduling process enabled the construction of a feasible schedule in a short CPU time. But this method may also have inefficient resource allocation because a subset of activities of projects was rescheduled for each number of iterations. So in the MIP based heuristic method, several activities were considered consecutively in each rescheduling process. Thereby a more efficient allocation of resources may be determined for project scheduling problem. The performance of rescheduling process is low in the MIP based heuristic method.

A scheme was proposed (Sharma, A., et al 2015) for the software project scheduling and allocation. In the proposed scheme Ant Colony Optimization and Event based scheduler techniques were utilized to develop a flexible and effective model for software project planning. The starting time and ending time of the project were considered as events in the Event Based Scheduler. The main objective of the event based scheduler was to adjust the allocation of employees at events and the unchanged events as non events. For the allocation of employees and planning Ant Colony Optimization was used. Initially the population of ant was initialized and each ant in the population scheduled the tasks to the employees based on their objective function. Finally, all the ants were showed the same project scheduling with better objective values. But the theoretical analysis of ACO is more difficult for software project selection and scheduling.

Two heuristic approaches is proposed (Waligóra, G. et al 2016) discrete-continuous project scheduling problems. It considered the convex processing rates functions of activities, maximization of NPV and positive discounted cash flows to solve the discrete-continuous project scheduling problems. The process of the first heuristic was to uniformly distribute the resources among the combination of employees and tasks which is called as HU. The process of the second heuristic was to increase the chance of activities with bigger cash flows to be completed earlier. Each activity requires for its processing discrete resources and an amount of a continuous resource.

A multi objective based Event based project scheduling using optimized Neural Network based Ant Colony Optimization (ACO) was proposed (Ponnam, V. S., & Geethanjali, N. 2015). In the conventional approach an event based scheduler Ant Colony Optimization was utilized on task scheduling. An optimized plan was developed from all iterations in the form of a matrix. Then the plans were developed and scheduled based on events. Thus, the ACO solved the problem of project scheduling. However, ACO does not consider the updated task allocation matrix for project scheduling process. Hence, in this paper, an improved ACO was proposed which optimized a global search through an introduced neural approach which scheduled the multi tasks in the project. Moreover, a multi objective approach was utilized to optimally schedule the number of tasks and resources involved in the software projects. When an uncertain event occurs, the remaining resources will be effectively computed, and the remaining tasks will be completed. A new schedule was created according to the calculation. An enhanced Entropy method can be utilized to represent the level of how much information or threshold has been figured out into the pheromone trails and subsequently, the heuristic parameter can be enhanced accordingly.

A multi objective evolutionary algorithm was developed (Hanne, T., and Nickel, S. 2005) for scheduling and inspection planning in software development projects. The problem of planning inspections and other tasks within the software development project with respect to the project duration, project costs and objective quality was considered in this paper. A simplified formulation of the problem as a multi objective evolutionary algorithm was developed comprising the phases such as test, coding, inspection and rework based on a discrete event simulation model of software development project. These problems were solved by finding an approximation of the efficient set in the software project through a multi objective evolutionary algorithm. But this method doesn't refine the behavioural model of the developers of the project which is main consideration of the stochastic nature of the software development process.

2.11.2 Resource Constrained Project Scheduling

A hybrid meta heuristics was proposed (Tritschler, M. et al 2017) for a flexible resource constrained project scheduling problem. The hybrid meta heuristics contains the Flexible Resource Profile Parallel Schedule Generation Scheme that utilized the concepts of non-greedy resource allocation and delayed scheduling, incorporated in a genetic algorithm. In addition to that, the best schedules were improved in a novel variable neighborhood search through transferring resource quantities between activities which were selected based on the analysis of resource flows. The main contributions of the proposed hybrid Meta heuristic was to embed a novel scheduling generation scheme into a genetic algorithm and enhanced the scheduling process in a variable neighborhood search and generated near optimal schedules in a short computation time.

An adaptive robust optimization model was proposed (Bruni, M. E. et al 2016) to address the problem of resource constrained project scheduling problem. This model derived the resource allocation decisions which minimized the worst case makespan under general polyhedral uncertainty sets. The properties of the proposed model were analyzed and it was considered that activity duration was subjected to interval uncertainty. In the intervals of uncertainty the level of robustness was controlled by a protection factor associated to the risk aversion of the decision maker. The counterpart of the resource constrained project scheduling problem was solved by the general decomposition approach. In addition to that, this approach was tailored to address the uncertainty set with the projection factor. But in this proposed model, ellipsoidal uncertainty sets were not considered.

A mixed integer linear programming model was proposed (Naber, A. 2017) for flexible resource constraint project scheduling problem. This model utilized a continuous time system to synchronize resources and activities. It can be achieved by the modified best discrete time model FP-DT3 model and forward and backward calculations. Then a simple left compression algorithm was processed on discrete time heuristic solutions. From this process, an upper bound of the continuous makespan was obtained. In addition to that, an event estimation method was proposed, which suggested the number of time intervals or time and mapped the earliest and latest time parameters on to the earliest and latest event parameters which were utilized in the proposed model. The computational speed of the proposed model was improved by certain families of valid inequalities; heuristic conditions and optimally based cuts were proposed as cuts in the branch and cut procedure. The efficiency of the proposed model is less.

A multi start iterative search heuristic approach was proposed (Zhu, X. et al 2017) for resource constrained project scheduling problem. The sequencing problem was solved by iterative search framework. It effectively finds out the activity sequences of the tasks. In the software project scheduling, the resource decision problem was resolved by introducing a backward peak elimination algorithm and two stage resource adjustment algorithms. The resource availability, cost problem was solved in the proposed approach by decomposing the problem into sub problems such as resource sequencing problem and decision problem were addressed consecutively. A feasible neighborhood and a path re-linking based method were developed to solve the sequential problem of the software project scheduling. Two Stage Resource Adjustment Procedure (TSRAP) and Backward Peak Elimination Procedure (BPEP) were constructed to solve the resource decision problem. The major drawback of this approach is that the time consumption is still high.

A dedicated local search method called as NALS method was proposed (Niño, K. et al 2016) for the multi objective resource constrained project scheduling problem. The primary intention of this NALS method was to reduce the total weighted start time and maximum completion time. The resource constrained project scheduling problem consists in scheduling a set of activities and the allocation of a set of limited resources with the consideration to optimize more than a single objective. Proposed method consists of two phases. First phase was focused on the minimization of makespan followed by the second phase which considered the improvement of total weighted start time. Set of solutions determined by NALS has a good coverage of the solution space. A set of non dominated solutions i.e., Pareto Front was obtained which offered the best solutions for both objectives. In addition to that, it balances the two objectives of the proposed approach with the intention of analysis by the decision maker quickly.

An integrated mixed integer programming model of generalized resource constrained multi project scheduling problem with a forward backward supply chain planning model was proposed (Gholizadeh-Tayyar, S. et al 2016). It described a transportation production plan to supply the non-renewable resources at the projects worksites and to ship the non-renewable resources to the recycling centers. In addition to that, a plan was defined to assign the renewable resources to the activities. By renting the capability of resources make the proposed model to obtain more optimum results in terms of paying for rent costs or penalty costs. But this proposed model was not applicable to large scale projects.

A model was proposed (Gutjahr, W. J. 2015) for stochastic multi mode resource constrained project scheduling problem under risk aversion. This model considered two objectives, which were cost and makespan. The activity cost and duration were considered as uncertain, and they were modelled as random variables. The class of early start policies was considered for the scheduling process of the decision problem. Moreover, along with the schedule, the assignment of execution modes to activities had been selected. For the resulting bi-objective stochastic integer programming problem the Pareto frontier was found through an appropriate solution method. It combined a branch and bound technique based on the forbidden set branching scheme from stochastic project scheduling.

A population based Meta heuristics algorithm called as discrete cuckoo search algorithm was proposed (Bibiks, K. et al 2015) to solve the resource constrained project scheduling problem. It was a modification of the cuckoo search algorithm which was adopted for solving the optimization problem. The core elements of cuckoo search algorithm were changed in the discrete cuckoo search algorithm. The changed core elements are objective function, egg, nest and movement through the search space which was used to solve the resource constrained project scheduling problem. But this approach failed to solve the more complicated scheduling problems.

A closed loop approximate dynamic programming was proposed (Li, H., and Womer, N. K. 2015) to solve the stochastic resource constrained project scheduling problem. The proposed efficient and approximate dynamic programming algorithm was based on the roll out policy. The heuristic algorithm was employed to improve the performance of roll out policy that improved the base policy offered by a priority rule heuristic. In addition to that, a hybrid approximate dynamic programming is proposed that integrated look ahead and look back approximation architectures that sequentially attained both the quality of look ahead and look back that sequentially enhance the task schedule process and efficiency of look back approach. Thus, the proposed hybrid approximate dynamic programming approach tackles the stochastic resource constrained project scheduling problem effectively.

2.11.3 Optimization Methods for Software Project Scheduling

An Ant Colony Optimization based Software Development Project Scheduling Problem methodology was proposed (Suri, B., and Jajoria, P. 2013) to solve the project scheduling problem. It returns results that can be obtained in the close proximity to the optimal results. The proposed methodology assisted the project managers to allocate the projects to the valuable employees with the objectives of minimum cost and less time duration of the project. The main intuition behind this approach was to provide a simplistic approach which solved the scheduling problem with minimum cost and minimum time consumption through Ant Colony Optimization algorithm. In this approach, only a few parameters were considered for software project scheduling.

An algorithm called Ant Colony Optimization Software Project Scheduling Problem (ACS-SPSP) (Xiao, J. et al 2013) employed ACO for the Software Project Scheduling Problem. Initially, the tasks were split and distributed employee dedications to task nodes. Then a construction graph was built, where the Software Project Scheduling Problem was converted into a graph based search problem. The ACO mechanism supported the use of domain knowledge as heuristics to improve the ant search ability process. Six heuristics were used in ACS-SPSP along with the allocated dedications of employees to other tasks, importance of the task in the project, the total dedications in tasks, a constant and both the total dedications in tasks. These are considered as the various aspects of the tradeoffs in SPSP to enhance the performance of software scheduling process.

A hybrid technique was proposed (Avinash, A., and Ramani, K. 2014) for Human Resource allocation and software project scheduling. In this proposed technique, task identification and task scheduling were implemented using Event Based Scheduler (EBS) and the efficiency of the allocation of tasks and employees were computed through Ant Colony Optimization (ACO) technique. In EBS based task scheduling, the commencing and ending of tasks were considered as events and it guaranteed optimized resource utilization. Through task precedence graph allocate the most suitable employee for the given project tasks. Task precedence graph was obtained through ACO, where the shortest path was found. The search space for employee allocation and the cost of the project was reduced by combining EBS and ACO. Thus, this proposed hybrid technique reduce the cost of software development and it readjusts the tasks which cannot be completed within the given deadline, which was based on task precedence graph. It also finds out the tasks priorities to adjust task precedence graph for task accomplishment. In ACO, the sequence of random decisions sometimes may affect the project scheduling performance.

Three cuckoo search based multi objective mechanisms were developed (Maghsoudlou, H. et al 2017) to concurrently optimize the conflicting objectives of minimizing the reworking risk and project cost of software. The developed algorithm was based on multi objective particle swarm, non dominance genetic algorithm and multi objective invasive weeds optimization algorithm. The Tauguchi method was applied to tune the parameters of the developed algorithm which improved the efficiency of the solution procedures. The performance of the proposed methodologies was computed by using a competitive multi objective invasive weeds optimization algorithm. The developed algorithms were compared in terms of metrics by using priority based methods. Thus, these developed algorithms enabled the project managers to effectively schedule the workforce of software projects in an appropriate way. The mixed combination of resources may give rise to some problems in software project scheduling which is not solved in this paper.

A modified grey shuffled frog leaping algorithm was proposed (Amirian, H., and Sahraeian, R. 2017) to optimize the problem of projection selection scheduling. The project selection scheduling problem was solved by using modified metaheuristics. Unlike deterministic parameters, gray parameters follow specific relation with constraints and also with each other. The bi level problems such as project selection scheduling searching on both levels utilized discrete operators such as shift, which improved the performance of an algorithm. This algorithm has high run time to add more realistic assumptions to the project model which was the major disadvantage of this algorithm.

The application of Particle Swarm optimization was investigated (Gerasimou, S. et al 2012) to create optimal project through specifying the best sequence for executing the project tasks and reducing the project duration. The presented approach helped the project managers perform the project scheduling and effective team staffing. In addition to that, it formed the productive and skillful project teams with the best utilization of developer skills. These requirements were encoded into the particle swarm optimization algorithm with different objective functions and hard constraints which generated the best solution with respect to their feasibility and the quality of the solution. The initial population of particle swarm optimization was initialized with the number of tasks and the employees and each particle in the population generate the best solution based on their objective function and constraints. But the particle swarm optimization has low convergence problem.

A quality based software project scheduling and staffing was proposed (Seo, D. et al 2015) based on a genetic algorithm. A quality score was defined which considered the practical issues in software project planning and it also considered the task severity and defect amplification model.

Moreover, in the genetic algorithm, cost was utilized as cost-bound which considered cost along with the quality. The software project was scheduled based on the consideration of employee, tasks and the cost along with the help of Genetic algorithm. The relationship between the tasks and the employees was described as the quality score of software product in terms of software defects. Quality score and cost of the project was assigned as objective function of the genetic algorithm. Genetic algorithm scheduled the tasks to the appropriate employees based on fitness values. The genetic algorithm cannot find the exact solution always for the task scheduling process, but it finds the best solution for task scheduling process.

The overall summary of the above techniques are listed in Table 2.1.

Table 2.1   Comparison of Different Software Project Scheduling

Author Name & Year	Methods	Demerits	Metrics
Shen, X. et al. 2016	Multi Objective Evolutionary Algorithm for dynamic project scheduling	Simplifications of this approach were not valid.	Project cost= 304597, Project duration= 5.30, Robustness= 0.12, Stability= 6.14
Stylianou, C. et al. 2012	IntelliSPM which has Multi Objective Genetic Algorithm followed by Single Objective Particle Swarm Optimization or Genetic Algorithm	This tool offered limited ability for project managers to have control over the objectives.	Project duration= 24days
Yannibelli, V. and Amandi, A. 2011	Knowledge based Evolutionary Approach	Complexity was high.	Crossover probability= 0.8, Mutation probability=0.05
Chang, C. K. et al. 2008	Time line based model with new Genetic Algorithm	Requires better representation of skills, employees and tasks.	Best cost= 22528
Padberg, F. and Weiss, D. 2011	Reinforcement learning based optimal policy	Less effectiveness.	Project cost= 9.8

Author Name & Year	Methods	Demerits	Metrics
Chen, W. N. and Zhang, J. 2013	Ant Colony Optimization with an Event based Scheduler	The human resources were not effectively managed.	Project cost= 2550000
Minku, L. L. et al. 2014	Improved Evolutionary Algorithm	Requires multi objective optimization algorithm.	Project completion time= 98, Average cost= 49998
Khodakarami, V. et al. 2007	Critical Path Method using Bayesian approach	Requires handling management actions for dynamic parameter learning.	Nil
Chicano, F. et al. 2011	Multi-objective Meta heuristics algorithm	Requires the most effective algorithm for software project scheduling.	Nil
Stylianou, C. and Andreou, A. 2011	Genetic algorithm for intelligent software project scheduling	Local optima & convergence problem in the GA affects the determination of optimal solution.	Project duration= 24days
Amiri, M. & Barbin, J. P. 2015	Differential Evolution algorithm	The average execution time and project costs were not optimized.	Average project duration= 21.88, Average project cost= 45230.15

Author Name & Year	Methods	Demerits	Metrics
Ren, J. et al. 2011	Cooperative Co-Evolutionary Algorithm	It has the problem of communication overhead and schedule robustness.	Project duration= 23.9
Mathew, J. et al. 2016	Multi Objective Optimization using genetic algorithm	Complexity was high.	Project duration= 106.4865days, Project cost= 1316773
Huang, X. et al. 2016	Mean variance and mean semi variance models	This model failed to take more life setting into account to solve the optimal project selection and scheduling problem.	Error rate= 0.6%
Xu, J. et al. 2015	New bilevel model with multiple decision makers and PSO	PSO has some convergence problem.	Nil
Zimmermann, A. 2016	Mixed Integer linear Programming	High complexity.	Nil
Sharma, A., et al. 2015	Event based scheduler and ant colony optimization	The theoretical analysis of ACO is more difficult.	Project cost= 2428000
Waligóra, G. et al. 2016	Two heuristic approaches for discrete-continuous project scheduling	Efficiency was less.	Average relative deviation= 10.55%, CPU time= 0.213sec

Author Name & Year	Methods	Demerits	Metrics
Ponnam, V. S. & Geethanjali, N. 2015	Multi objective based Event based project scheduling using optimized Neural Network based Ant Colony Optimization	Convergence problem emerged.	Scheduling accuracy= 0.96, Average deviation= 0.98
Hanne, T. & Nickel, S. 2005	Multi objective evolutionary algorithm and inspection planning	This method doesn't refine the behavioral model of the developers of the project.	Average duration= 6.87
Tritschler, M. et al. 2017	Hybrid meta heuristics	Requires optimal allocation of resources.	Average scheduling time= 1.15sec
Bruni, M. E., et al. 2016	Adaptive robust optimization model	Ellipsoidal uncertainty sets were not considered	Makespan= 81.73, Time= 118.78sec
Naber, A. 2017	Mixed integer linear programming model	Efficiency was less.	Average makespan= 46.13, Average CPU time= 14sec
Zhu, X. et al. 2017	Multi start iterative search heuristic approach	The time consumption of the proposed approach is still high.	CPU time= 39sec
Niño, K. et al 2016	Dedicated local search method	Efficiency was less.	Nil

Author Name & Year	Methods	Demerits	Metrics
Gholizadeh-Tayyar, S. et al 2016	Integrated mixed integer programming model	This proposed model was not applicable to large sizes of projects.	Nil
Gutjahr, W. J. 2015	Bi-ObjectiveMulti-ModeProjectScheduling	Complexity was high.	Makespan= 465.650, Cost= 576.705
Bibiks, K. et al 2015	Discrete cuckoo search algorithm	This approach failed to solve the more complicated scheduling problems.	Average deviation= 0.44
Li, H., and Womer, N. K. 2015	Closed loop approximate dynamic programming	Computation complexity was high.	Average CPU time= 0.070sec
Suri, B., and Jajoria, P. 2013	Ant Colony Optimization based Software Development Project Scheduling Problem	Only a few parameters were considered	Project completion time= 18days
Xiao, J. et al. 2013	Ant Colony Optimization Software Project Scheduling Problem	Multi software project scheduling problem was not investigated.	Duration= 214369, Cost= 1200000
Avinash, A., and Ramani, K. 2014	Hybrid technique using Event Based Scheduler and ACO	The sequence of random decisions in ACO affects the performance.	Efficiency= 93%

Author Name & Year	Methods	Demerits	Metrics
Maghsoudlou, H. et al 2017	Three cuckoo search based multi objective mechanisms	The mixed combination of resources may give arise to some problems in software project scheduling.	Average mean ideal distance= 43.10, Average spacing metric= 0.0199
Amirian, H. & Sahraeian, R. 2017	Modified grey shuffled frog leaping algorithm	High run time.	Error rate= 0.85±0.02
Gerasimou, S. et al 2012	Particle Swarm optimization	Low convergence problem occurred.	Average feasibility rate= 33.1%, Hit ratio= 89%
Seo, D. et al 2015	Quality based software project scheduling by genetic algorithm	Less performance.	Quality score= 0.85

2.12 SUMMARY

In this chapter, the principles of software project management, task scheduling and resource allocation were discussed in detail. Project planning and scheduling are acquainted with a extensive range of engineering domains. Both making plans and scheduling are extensively essential and Planning also consists of scheduling on account that decreasing delays during individual jobs allows supervisors to assign greater jobs. Scheduling solutions the question of what number of jobs to assign, while the making plans concepts deal with foremost problems, the scheduling principles are more of a framework. Project Evaluation and Review Technique (PERT) specifies technological precedence constraints amongst duties in a project. Work breakdown structure (WBS) organizes obligations into purposeful organizations. The Critical Path Method (CPM) assumes endless resources, which if not immediately available, can be acquired at some cost. When useful resource constraints are introduced, the problem will become plenty greater tough to resolve. And in maximum cases, the greatest answer is unknown. This is pretty frequently legitimate because the number of tasks and quantity of resources increases. Moreover in this chapter, the previous researchers associated to the software project task scheduling problem in software project management environment were discussed in detail. The software project task scheduling problem was solved by different techniques and approaches which were perused in detail. This study is useful for identifying the challenges involved in software project task scheduling process and based on these challenges, the proposed solution for project task scheduling problem is presented.

CHAPTER 3

AN OPTIMIZED EVENT BASED SOFTWARE PROJECT SCHEDULING WITH UNCERTAINTY

3.1 INTRODUCTION

Software project planning plays the most important role in the end product of software. To plan the software project, software project managers takes responsibility. During planning the software project, resource allocation and project scheduling are the most significant issues. Scheduling refers the process used for ordering the tasks in the project in order to execute within the given time. In addition, resource allocation is also the process used for assigning the available resources to the tasks for its execution. Generally, both scheduling and allocation are having high complexity due to the considered constraints and factors in the project. The software construction cost can be reduced by an efficient project plan for providing the successes to the organizations.

Over the past decades, the problems in the task scheduling also known as NP complete have been resolved by different approaches. The project scheduling process is directed regularly by the software project managers in order to perform preface time and estimate the resources, general assistance and analyze the project alternatives. The challenging task in planning is using computer software for software projects due to the most sensitive resources and activities. For scheduling and human resource allocation, various software modelling techniques have been presented. However, the variations in combining and leaving the events of human resources have not considered in many of the modelling techniques. These issues have been avoided by an event based Ant Colony Algorithm (ACA) (Chen, W. N., & Zhang, J. 2013). In this approach, these issues can be considered into account by representation of the task list and a planned employee allocation matrix. The reconstruction of allocation of employee's at events and maintenance of the unchanged allocation at non-events are the fundamental objective of EBS. In software project management, a significant factor is an uncertainty event which may cause the failure of the current plan. An uncertainty event is raised from knowledge dependencies like contextual information about the project, considering the underlying process, previous events explanation and modified speed. Hence, in this research work, the issues and solutions of an uncertain event are addressed and provided which may improves the software project planning and resource allocation.

3.2 DESCRIPTION OF EMPLOYEES

To categorize the information of employee's like salaries, skills, and working constraints, employee's database is required. In employee allocation process, the most considerable issue is assigning the employees to suitable tasks which help to complete the tasks efficiently. Considering that m employees are involved in the project, the following attributes are considered for the ith employee (i-=1,2,…,m).

- b_{si}: Employee's basic salary per time period.
- h_{si}: Employee's salary per-hour for normal work.
- oh_{si}: Employee's salary per-hour for over-time work.
- nh: Legal normal working hours per month.
- $maxh_i$: Employee's maximum possible working hours per month for the project.

- $[join_i, leave_i]$: The time window when the employee is available for the project.
- $\{s_i^1, s_i^2, \ldots, s_i^{\Phi}\}$: Employee's skill list, where Φ is the number of skills and $s_i^j = 0$ means that an employee does not have any skill and $s_i^j = 5$ means that an employee has mastered on that skill.
- $taskexp_e$: Employee's experience level.

If employees work on a requiring task with low proficiency on the skill list, then they will become more proficient as a result of their experience. This refers on-the-job training. It is assumed that for each skill area, each employee has an initial experience level. At the primary time period, an employee's experience level on a task is r and they work for a fraction of time (b) on the task, then at the end of the time period their experience will be $taskexp_e = \max\left(join_i + \frac{\mu}{\Phi} \times b, 5\right)$. Here, μ is the employee's learning speed. The maximum value for experience is similar as that for proficiency.

Assume $Lthr$ is the leaning speed threshold value for employee, nls is the normal learning speed value and μ is the employee learning speed. For i[th] the employee at t[th] month, $salary_i^t$ (Chen, W. N., & Zhang, J. 2013) is determined by the following equation.

$$salary_i^t = \begin{cases} b_{si} + hours_i^t . h_{si} + \mu, & hours_i^t \leq nh, \mu \leq Lthr \\ b_{si} + nh. h_{si} + (hours_i^t - nh). oh_{si}, & nh < hours_i^t \leq maxh_i, nls < \mu \leq Lthr \\ \infty, & hours_i^t > maxh_i, \mu > Lthr \text{ 0therwise} \end{cases}$$

(3.1)

3.2.1 DESCRIPTION OF TASKS

The task can be described by using the Task Precedence Graph (TPG). A TPG is an acyclic directed graph G(T,A). The set of nodes $T\{t_{p1}, t_{p2}, \dots, t_{pn}\}$ corresponds to the set of tasks, where n is the number of tasks in the project. For a task t_j ($j = 1, 2, \dots, n$), considered the following attributes:

- pm_j: Estimated work effort of the task is in months. Different methods are used for estimating the work effort.
- SK_j: Set of skills required by the task.
- $maxhead_j$: Maximum headcount for the task.
- $deadline_j$ and $penalty_j$: Deadline and penalty of the task. The penalty will be collected for the delayed task.

According to these attributes, consider wh_{ij}^t refers to the amount of working hours of the ith employee for t_j at the tth month. The achievement A_j^t yielded by the employees for t_j at time t is computed by using the following steps.

- The proficiency prof$_{ij}$ of the ith employee for t_j is follows,

$$prof_{ij} = \prod_{id \in SK_i} s_i^{id}/5 \qquad (3.2)$$

- An experience level on a task for employee i's is join$_i$ at the beginning of a time period and they work for a fraction of a time (b) on this task is as follows,

$$Taskexp_e = \max\left(join_i + \frac{\mu}{\phi} \times b, 5\right) \qquad (3.3)$$

- The employees overall fitness f_j^t for t_j on the tth month is evaluated by the following equation,

$$f_j^t = \frac{\sum_{i=1}^{m} prof_{ij}.wh_{ij}^t}{\sum_{i=1}^{m} wh_{ij}^t} + \frac{\sum_{i=1}^{m} taskexp_e.weh_{ij}^t}{\sum_{i=1}^{m} weh_{ij}^t} \qquad (3.4)$$

- Transform f_j^t to a cost driver value $V = 8 - round(f_j^t.7 + 0:5)$, where the value of V belongs to 1 to 7. The most suitable employees for the task and vice versa are referred by V=1.
- Then, the achievement A_j^t for t_j on the tth month is determined as follows,

$$A_j^t = \frac{\sum_{i=1}^{m} wh_{ij}^t}{V} \qquad (3.5)$$

3.3 PLANNING OBJECTIVE FUNCTION

The specification of when the tasks of the project are processed and how the employee's workloads are allocated to the task are done by using a plan for the project. Especially, the plan has the start time ($start_j$) and the finish time ($finsh_j$) of each task$_j$ ($j \in \{1,2,...,n\}$) and the working hours $\left(wh_{ij}^t\right)$ of all employees $i \in \{1,2,...,m\}$ to the task t_j during the time window $t = \{start_j, finish_j\}$. The following constraints should satisfy by the plan:

- The working hours of ith employee per month should not exceed the limit $maxh_i$ such as given below.

$$\sum_{j=1}^{n} wh_{ij}^t = hours_i^t \leq maxh_i, t = 1 \qquad (3.6)$$

- The working hours of ith experienced employee learning speed must not exceed the limit.

$$\sum_{j=1}^{n} \mu = weh_{ij}^t \leq Lthr, t = 1 \qquad (3.7)$$

- The number of employees apportioned to the task t_j is limited by the maximum headcount such that,

$$\sum_{i=1}^{n} sign\left(\sum_{t=start_j}^{finish_j} wh_{ij}^t\right) \leq maxhead_j \quad (3.8)$$

$$\text{Where } sign(x) = \begin{cases} 1 \text{ if } x > 0 \\ 0 \text{ if } x = 0 \end{cases}$$

- All tasks have to be complete. In other words, for a task t_j, the sum of the achievements for t_j during the time window (start$_j$, finsh$_j$) must fulfill,

$$\sum_{t=start_j}^{finish_j} A_j^t \geq pm_j \quad (3.9)$$

This approach deliberates cost minimization as the objective function which is given in following equation

$$minf = \sum_{t=1}^{end} salary_i^t + \sum_{j=1}^{n} penalty_j + \sum_{j=1}^{n} \mu \quad (3.10)$$

3.4 EVENT-BASED SCHEDULER MODEL

This offers a representation scheme with the novel event-based scheduler. The task list is represented as $(t_{p1}, t_{p2}, \ldots, t_{pn})$ and the planned employee allocation matrix is represented as,

$$\begin{Bmatrix} pwh_{11}pwh_{12} \ldots pwh_{1n} \\ pwh_{21}pwh_{22} \ldots pwh_{2n} \\ \vdots \quad \vdots \quad \ddots \quad \vdots \\ pwh_{m1}pwh_{m2} \ldots pwh_{mn} \end{Bmatrix}$$

The planned experienced employee allocation matrix is represented as,

$$\begin{Bmatrix} weh_{11}weh_{12} \ldots weh_{1n} \\ weh_{21}weh_{22} \ldots weh_{2n} \\ \vdots \quad \vdots \quad \ddots \quad \vdots \\ weh_{m1}weh_{m2} \ldots weh_{mn} \end{Bmatrix} \quad (3.11)$$

The Event Based Scheduler (EBS) amends a plan in the form of equation (3.11) into an actual timetable by two rules. First, if there is a resource conflict between two tasks, the task that appears earlier in the task list has the highest priority to use the resource. That refers, presuming that the ith employee is formerly stated to concurrently dedicate pwh_{ij} and pwh_{ik} of his working hours to t$_j$ and t$_k$ respectively, if $pwh_{ij} + pwh_{ik} > maxh_j, maxh_i$, the employee will first dedicate his working hours to the task with the highest priority, $weh_{ij} + weh_{ik} = Lthr$. Second, new workload assignments are only made while events occur. If no employee joins or leaves the project or no human resource is released by the tasks just finished, the workload assignments remain the same as the previous time period.

3.5 ANT COLONY OPTIMIZATION ALGORITHM

In this section, ACO algorithm for task scheduling is discussed in brief. Initially, the task list is constructed. A task list is an order of tasks $(t_{p1}, t_{p2}, \ldots, t_{pn})$ which satisfies the precedence constraints defined by the TPG. Here, the pheromone and the heuristic are defined for task list construction. Pheromone is used for constructing the task list in which an order of the tasks is determined by an ant. For the considered software project planning, only one employee may undertake the different tasks simultaneously with considered skill proficiency. It has more complex for defining the related tasks for the relation-learning model since only one task is assigned to various employees. Hence, an absolution position model is introduced with the summation rule.

3.5.1 Construction of Task List

The Minimum Slack (MINSLK) heuristic is adopted for task list construction. A task with a relatively smaller MINSLK implies that this task is more imperative. Initially for each task, the shortest possible makespan is estimated. Then, the earliest start time and the latest start time of each task are evaluated based on the computed the shortest possible makespan of each task.

After that, the MINSLK is determined according to the difference between the latest start time and the earliest start time of the task.

For constructing the feasible task list, each ant maintains an eligible set of the tasks that satisfy the precedence constraint. The following steps are kept in mind while constructing a task list.

- Set the tasks that can be implemented at the beginning of the project, such that the tasks which do not have any precedence tasks fall into the eligible set.
- For k=1 to n, do the following processes continuously.
- Select the task from eligible set and put the task to k^{th} position of the task list.

$$\Pr(j,k) = \begin{cases} \frac{\sum_{l=1}^{k} \tau_t(j,l) \cdot \eta_t(j)}{\sum_{t_u \in eligibleSet} \sum_{l=1}^{k} \tau_t(u,l) \cdot \eta_t(u)}, & if\ t_j \in eligibleSet \\ 0, & Otherwise \end{cases} \quad (3.12)$$

$$\eta_t(j) = 1/MINSLK_j \quad (3.13)$$

Here, $\eta_t(j)$ is called as the heuristic for task t_j and $\Pr(j,k)$ is the probability of selecting the task t_j. The pheromone of putting a task t_j to k^{th} position of the task list is denoted as $\tau_t(j,k)$

- Update the eligible set by removal of the selected task from eligible set and include the new feasible tasks which satisfy the precedence constraint into eligible set.
- After much iteration, a feasible task list is constructed.

3.5.2 Construction of Employee Allocation Matrix

The employee allocation matrix was used for specifying the originally planned working hours of employees to tasks. For constructing the employee allocation matrix, two types of pheromone are adopted. The pheromone of selecting i^{th} employee to work for the task t_j is denoted as $\tau_{e1}(u,j)$. The heuristic of selecting i^{th} employee to work for the task t_j is denoted as $\eta_e(i,j)$.

$$\eta_e(i,j) = \frac{prof_{ij} \cdot askexp_e}{hs_i} \qquad (3.14)$$

The employee allocation matrix is constructed based on the following steps:

- Set all values in the employee allocation matrix to 0.
- For each task t_j ($j = 1,2,\dots,n$), assign the workloads for t_j by following steps.
- Evaluate the value of $\tau_{e1}(i,j) \cdot \eta_e(i,j)^\beta$ for all employees where β is a parameter.
- Select the employee from eligible set Eligible (Chen, W. N., & Zhang, J. 2013) and set the task to k^{th} position of the employee allocation matrix.

$$\Pr(u,j) = \begin{cases} \frac{\tau_{e1}(u,j) \cdot \eta_e(i,j)^\beta}{\sum_{t_u \in Eligible} \tau_{e1}(u,j) \cdot \eta_e(i,j)^\beta}, & if\ u \in Eligible \\ 0, & Otherwise \end{cases} \qquad (3.15)$$

- Update the eligible set by removal of the selected employee from eligible.
- Assign the working hours for the employee selected to work for t_j by using pseudorandom proportional rule. The probability of assigning the working hours is given by,

$$\Pr(i,j,k) = \begin{cases} \frac{\tau_{e1}(i,j,k)}{\sum_{v \in \{25\%nh, 50\%nh, \dots, maxh_i\}} \tau_{e2}(i,j,v)} \end{cases} \qquad (3.16)$$

$$k \in \{25\%nh, 50\%nh, \dots, maxh_i\}$$

- Repeat the process until the workload assignment process is completed and then finally the employee allocation matrix is constructed.

3.5.3 Pheromone Management

All the pheromone values are set to an initial value $\tau_{initial}$ and is expressed as follows,

$$\tau_{initial} = \frac{1.0}{\sum_{i=1}^{m}[b_{si}+maxsalary_i]\cdot(deadline+n)} \quad (3.17)$$

Here, $maxsalary_i$ is the maximum possible salary of ith employee per time period (Chen, W. N., & Zhang, J. 2013) which is calculated by following equation.

$$maxsalary_i = \begin{cases} b_{si} + maxh_i \cdot h_{si} + \mu, & 0 \leq maxh_i \leq nh, \\ b_{si} + nh \cdot h_{si} + (maxh_i - nh) \cdot oh_{si}, & maxh_i > nh. \end{cases} \quad (3.18)$$

The upper bound is estimated by considering that all employees contribute all of their working hours to the project during each time period and the project is delayed by 'n' time periods with respect to the predefined deadline. During the construction of task list and employee allocation matrix, the local pheromone updating rule is applied in reducing the corresponding pheromone values for selecting the other possible solutions. If the task t_j is selected to k^{th} position of the task list, the corresponding pheromone is updated as,

$$\tau_t(j,k) = (1-\rho) \cdot \tau_t(j,k) + \rho \cdot \tau_{initial} \quad (3.19)$$

Here, ρ is a parameter. Similarly, if i^{th} employee is assigned to j^{th} task with k working hours, then the corresponding pheromone is updated as,

$$\tau_{e1}(i,j) = (1-\rho) \cdot \tau_{e1}(i,j) + \rho \cdot \tau_{initial} \quad (3.20)$$

$$\tau_{e2}(i,j,k) = (1-\rho)\cdot\tau_{e2}(i,j,k) + \rho\cdot\tau_{initial} \qquad (3.21)$$

At the end of the iteration, an additional pheromone is deposited to the components which are associated with the best-so-far plans discovered by the algorithm. Therefore, all pheromone values associated with the best-so-far plan are updated by,

$$\tau_t(j,k) = (1-\rho)\cdot\tau_t(j,k) + \rho\cdot\Delta\tau \qquad (3.22)$$

$$\tau_{e1}(i,j) = (1-\rho)\cdot\tau_{e1}(i,j) + \rho\cdot\Delta\tau \qquad (3.23)$$

$$\tau_{e2}(i,j,k) = (1-\rho)\cdot\tau_{e2}(i,j,k) + \rho\cdot\Delta\tau \qquad (3.24)$$

Here $\Delta\tau = 1/f$ is the reciprocal of the objective function value of the plan.

3.5.4 Local Mutation Process

Moreover, ACO algorithm is inspired by the mutation operator which is used by genetic algorithm. The following steps and operators are used for mutating the best-so-far solution which is discovered by ACO.

- Operator 1-Task list mutation: Initially, a task is randomly selected from the task list. Then, the destination position is randomly generated. The mutation is performed by moving the selected task towards the destination position in the task list.
- Operator 2-Employer allocation matrix mutation: Initially, a task is selected randomly. Then, an employee who has been assigned to work for this task is randomly selected and his working hours for this week are reset. Finally, a new employee who has not been selected to work for the task is randomly selected and his working hours for the task are randomly set to the value.

During all iterations, mutatenum of mutated solutions are generated based on either of the above mutation operators. If any mutated solution is better than the current best-so-far solution, then the best-so-far solution will be replaced. Then, a new simulation model is developed which constructs on the experience gained in our initial implementation.

3.6 RESULTS AND DISCUSSION

In this section, the experimental results of the proposed approach are illustrated.

3.6.1 Dataset Description

The Dataset of the work taken from Project Scheduling Problem Library (PSBLIB) (Kolisch, R., & Sprecher, A. 1997) was tested in this experiment. The four kinds of personnel in a group are the regular chosen staff, regular usual staff, and temporary skilful and temporary usual staff. The chosen staffs are the ones who have more skills and the employer has to pay a huge sum for them. The Usual staffs are also skilful, but they are not well versed in all aspects. The Temporary staffs are equal to the usual staff, but their services will be required only for some specific jobs, and they will be sent back once their work is done.

3.6.2 Experimental Setup

This experiment can be carried out on key information of the test instances which is mentioned in Table 3.1. Here the test instances taken for the tests are Project_1, J3010_1 to J3010_10, J6011_1 to J6011_10, J6030_1 to J6030_10, J9010_1 to J9010_10, and J9015_1 to J9015_10. These are assigned with the task number 11, 32, 62, 62, 92 and 92 respectively. Each and every skill is allotted an integer value and the employee number is an employee id which shows who is eligible to do the task on time.

Table 3.1 Information of the Test Projects

Name of the Project	Task Number	Skill Proficiency	Employee Id	Max number of solutions
Project_1	11	5	10	60000
J3010_1—J3010_10	32	5	10	100000
J6011_1—J6011_10	62	5	10	200000
J6030_1–J6030_10	62	10	20	200000
J9010_1–J9010_10	92	8	15	300000
J9015_1–J9015_10	92	8	15	300000

3.6.3 Project Completion Time versus Project Completion Cost

The objective function is minimizing the project completion cost and completion time of the software project. Hence, the performance of the proposed approach is analyzed with other methods in terms of cost and time. The following table 3.2 shows the comparison of Project Completion Time versus Project Completion Cost.

Table 3.2 Comparison of Project Completion Time versus Project Completion Cost

Project Completion Time (ms)	Project Completion Cost (Rs)			
	TS	KGA	ACO-L	ACO-UH

500	6100	5800	5400	5350
1000	6220	5980	5510	5425
1500	6370	6060	5620	5550
2000	6450	6280	5860	5692
2500	6600	6390	5940	5750
3000	6720	6500	6020	5850
3500	6800	6600	6100	5950
4000	6840	6710	6180	6050
4500	6900	6800	6240	6150

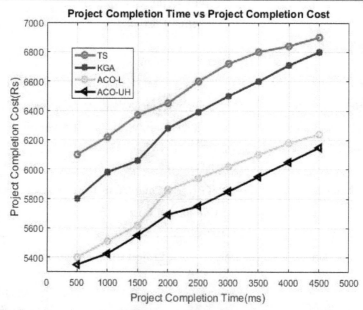

Figure 3.1 Comparison of Project Completion Time versus Project Completion Cost

Figure 3.1 shows that the comparison results of the proposed approach and existing approach in terms of cost and time parameters. The time taken to frame a schedule and resource allocation in the project is very low and the cost taken for scheduling and resource allocation is also very low in the proposed system. When the project completion time is 4500, the corresponding project completion cost of TS approach is 10.87%, KGA approach is 9.56% and ACO-L approach is 1.44% lesser than proposed ACO-UH approach.

3.6.4 Project Completion Time

The project completion time is defined as the amount of time taken to complete tasks successfully in a project. The following table 3.3 shows the comparison of project completion time.

Table 3.3 Comparison of Project Completion Time

	Methods			
	TS	KGA	ACO-L	ACO-UH
Project Completion Time (ms)	7000	6800	6500	6100

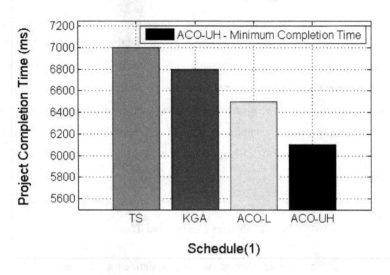

Figure 3.2 Comparison of Project Completion Time

Figure 3.2 shows that the comparison results of proposed approach and existing approach in terms of project completion time. The time taken to schedule and resource allocation in the project is very low in the proposed system.

3.6.5 Project Completion Cost

The project completion cost is the amount cost required for a project. The following table 3.4 shows the comparison of project completion cost.

Table 3.4   Comparison of Project Completion cost

	Methods			
	TS	KGA	ACO-L	ACO-UH
Project Completion Cost (Rs)	6650	6400	6140	6000

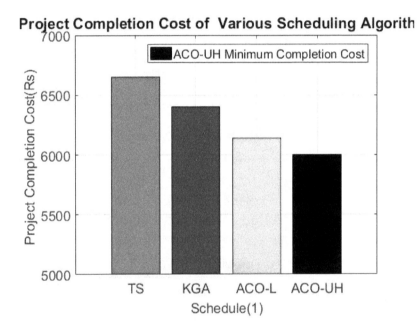

Figure 3.3   Comparison of Project Completion cost

Figure 3.3 shows that the comparison results of the proposed and existing approaches in terms of project completion cost. The cost to schedule and resource allocation in the project is very low taken for scheduling and resource allocation is very low in the proposed system.

3.6.6 Project Allocation Time

The project allocation time is defined as the amount of time taken by each method to schedule tasks of software projects. Here, the project allocation time is the amount of time taken by TS, KGA, ACO-L and ACO-UH approach for scheduling tasks of software projects. The following table 3.5 shows the comparison of project allocation time.

Table 3.5   Comparison of Project Allocation Time

	Methods			
	TS	KGA	ACO-L	ACO-UH
Project Allocation Time (ms)	6870	6500	6147	5421

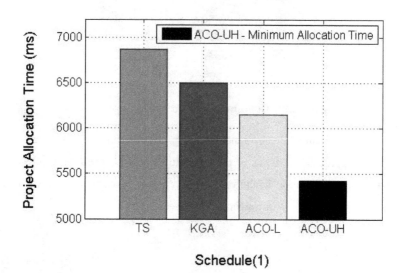

Figure 3.4   Comparison of Project Allocation Time

Figure 3.4 shows that the comparison results of the proposed and existing approaches in terms of project allocation time. The time taken to allocate the tasks to the appropriate employees by the proposed method is very low in the proposed system.

3.6.7   Productivity

The productivity of software is typically measured by the project size divided by effort. Effort is commonly expressed in man-months of labor. It can be mathematically expressed as follows:

$$Productivity = \frac{Number\ of\ function\ points}{Effort\ in\ man\ months}$$

The following Table 3.6 shows the comparison of productivity.

Table 3.6   Comparison of Productivity

	Methods			
	TS	KGA	ACO-L	ACO-UH
Productivity	80	85	88	90

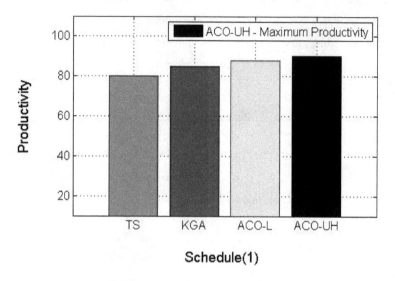

Figure 3.5   Comparison of Productivity

Figure 3.5 shows that the comparison results of proposed approach and existing approach in terms of productivity. The productivity of project scheduling and resource allocation is very high for the proposed approach.

3.7   SUMMARY

In this chapter, a novel method is proposed for solving the software project planning problem. In the proposed approach, there are two main approaches presented. At First, an event-based scheduler is introduced. Second, an advantage of the ACO is applied for solving the complicated planning problem. Better insights into the criticality of the tasks can help software development managers cope with the uncertainties that they face in project planning. The simulation with sufficiently accurate models can achieve better job of task estimation criticality than static analysis. In addition, a decision tool is developed that would help managers with better insights into the criticality of project tasks, as discovered by simulating the way the various uncertainties might unfold and interact as the project progresses. The experimental results show that the proposed ACO-UH improves the productivity such as 2.27% than ACO-L, 5.88% than KGA and 12.5% than TS techniques as well as reduces the project allocation time such as 11.81% than ACO-L, 16.6 than KGA and 21.09% than TS. In addition, this approach reduces the project completion cost such as 2.09% than ACO-L, 9.37% than KGA and 12.26% than TS techniques. Hence, the results prove that the proposed ACO-UH has better performance in terms of minimum cost and time compared to the other software project scheduling techniques.

CHAPTER 4

FUZZY C MEANS CLUSTERING WITH ACO FOR OPTIMIZED SCHEDULING AND RESOURCE ALLOCATION

This chapter provides the detailed information about the Fuzzy C means with ACO based project task scheduling and resource allocation for software project management. This chapter describes how the proposed Fuzzy C means with ACO based task scheduling and resource allocation increases the resource flexibility.

4.1 INTRODUCTION

In the project task scheduling, the project managers allocate the tasks to the most suitable employees based on the consideration different facts like cost, dead line and profit etc., the uncertainties due to the deficiencies in knowledge such as contextual information about the project, employees understanding about underlying process, speed of change and explanation of past events. In the previous work, these uncertainties were handled by the simulation tool which assigned tasks to suitable employees. But in this work, the flexibility of the human resources was decreased due to appointing the employee to only one task. In order to improve the resource flexibility, Fuzzy C means with Ant Colony Optimization was proposed. Initially, a representation scheme Fuzzy C Means (FCM) clustering methods (Azzeh, M., et al 2008) for Event Based Scheduler (EBS) is developed where the similar employee characteristics data are grouped in a cluster based on the experience time and task completion time of each one of the employee in the software projects. The representation scheme is composed of task list based on the employee experience for each task and a planned employee allocation matrix. The planned employee allocation matrix states the premeditated workload assignments. Finally, an ACO algorithm is used to solve the project scheduling problem, where the tasks are scheduled within the cluster based on the objective function. Thus, the search space of the ACO is reduced. The proposed Fuzzy C Means with ACO based task scheduling and resource allocation was tested in terms of cost and time.

4.2 FUZZY C MEANS WITH ACO BASED TASK SCHEDULING AND RESOURCE ALLOCATION

The EBS is an employee allocation and task list matrix where the both employee allocation and task scheduling problems were addressed. It adjusted the employee allocation whenever an event occurs. The employee allocation is keep as constant when the events do not take place. It schedules the tasks to the available resources to complete the software project. The EBS is represented as follows:

$$\begin{bmatrix} Task\ list: (t_{p_1}, t_{p_2}, \ldots, t_{p_n}) \\ Planned\ emplyee\ allocation\ matrix \begin{Bmatrix} pwh_{11} pwh_{12} pwh_{1n} \\ pwh_{21} pwh_{22} pwh_{2n} \\ \vdots \quad \vdots \quad \vdots \\ pwh_{m1} pwh_{m2} pwh_{mn} \end{Bmatrix} \end{bmatrix} (4.1)$$

In the above equation $t_{p_1}, t_{p_2}, \ldots, t_{p_n}$ denotes the task list, $p_1, p_2, \ldots p_n$ is a permutation of (1,2, ... n) and pwh denotes the planned working hours. The event based scheduler based on the clustering results is performed. The Clustering method clusters the employees based on their experience and task completion time. In this work, Fuzzy C means clustering is used to assemble the task list of the employees and the number of attributes of the employee based on their experience. The conventional FCM clustering method utilizes the distance measure to calculate the difference between the employees in the project but it ignores the global view clustering results. So in this proposed work, a regulatory factor based cluster density is used to measure the correspondence between different employee data objects. The regulatory factor based cluster density measure function is dynamically corrected by the regulator factor until the objective criterion is achieved. An employee is denoted as emp_i i=1,2,...n for every data emp_i, the dot density is written as follows:

$$Z_i = \frac{1}{\min\min(d_{ij})} d_{ij} \leq emp, 1 \leq i \leq n \quad (4.2)$$

In the above equation 4.2, d_{ij} represents the distance between the two employees attributes, attributes of employee is given as follows:

$$emp_1 = \{b_{si}, h_{si}, oh_{si}, nh, \max h_i, join_i, leave_i, s_i^1, s_i^2, \ldots s_i^\phi, Taskexp_e (4.3)$$

Every employee cluster density is the weighted combination of dot densities which is denoted as follows:

$$\hat{Z}_i = \frac{\sum_{j=1}^n a_{ij} w_{ij} z_{ij}}{\sum_{j=1}^n a_{ij} w_{ij}} d_{ij} \leq emp, 1 \leq i \leq n \quad (4.4)$$

In the above equation 4.4, a_{ij} represents the category is the label of employee data, w_{ij} denotes the weight of emp_j and w_{ij} is the positive constant that can be tuned by users. a_{ij} is assigned as 1 when emp_j is most likely fit to the cluster i otherwise it is assigned as 0. Using the cluster density Z_i is redefined as follows:

$$\hat{d}_{ij}^2 = \frac{||emp_j - v_i||^2}{\hat{z}_i} 1 \leq i \leq c, 1 \leq j \leq n \quad (4.5)$$

The optimization expression can be defined as follows in the equation 4.6,

$$J_{FCM-CD}(U,V) = \sum_{i=1}^{c}\sum_{j=1}^{n} u_{ij}^{m} ||emp_j - v_i||^2 \quad (4.6)$$

The Lagrange Multiplying method is applied in the equation 4.5, the two updated equations are obtained, which are given as follows:

$$v_i = \frac{\sum_{j=1}^{n} u_{ij}^{m} emp_j}{\sum_{j=1}^{n} u_{ij}^{m}} \quad 1 \leq i \leq c \quad (4.7)$$

$$u_{ij} = \frac{\hat{d}_{ij}^{-2/(3-1)}}{\sum_{j=1}^{n} \hat{d}_{ij}^{\frac{2}{m-1}}} \quad (4.8)$$

The metaphor of planning is the major difference between the existing and proposed scheme. In this work, the resource constrained project scheduling problem and employee assignment models plan with the metaphor of the task. Though this process makes plan with each task having the fixed workload assignment but it cannot deal with task pre-emption. Thus, this model is a fine- grained model and a workload assignment model and the large search space was reduced by using this model. The overall process of the FCM clustering is given as follows:

Fuzzy C Means Clustering algorithm

Input: Task list, employee list, employee attributes, task attributes

Output: Cluster of employee

Step 1: Select the number of cluster c, fuzziness index m, iteration error, maximum iterations T

Step 2: Initialize the membership degree matrix $U^{(0)}$

Step 3: Obtain the initial centroids using equation 4.8

Step 4: Compute the dot density of every employee data points using equation 4.4

Step 5: Update the membership degree matrix $U^{(t)}$ and cluster centroids $V^{(t)}$ using the equation 4.7 and 4.8 when the iteration index is t.

Step 6: Calculate the value of the objective function $J^{(t)}$ using equation 4.6

Step 7: If $|U^{(t)} - U^{(t+1)}| < a$ or t=T, then stop the iteration.

Step 8: Get the membership degree matrix and the cluster centroids V, otherwise set and return to step 5.

The clustered employees were used in the ACO (Monica, D., et al 2014) to schedule the task to the more capable employees. The basic idea of ACO is to motivate the foraging behavior of the ant colonies. Each ant in the population utilized a special kind of chemical called as pheromone to communicate with other ants. Each ant tries to allocate the tasks with the more appropriate employee. Initially, ants allocate the tasks randomly to the clustered employees. When an ant finds a better fitness value than the present fitness value, they leave pheromone on the path. An ant can follow the trails of the other ants to the schedule tasks by sensing the pheromone on the ground. As this process continues, most of the ants gets attracted to choose the shortest path, as there have been a huge amount of pheromones accumulated on this path. Each ant in the population updated their pheromone by using the following equation:

$$\tau_{xy} \leftarrow (1-\rho)\tau_{xy} + \sum_k \Delta\tau_{xy}^k \qquad (4.9)$$

In the above equation 5.9, τ_{xy} is the amount of pheromone deposited for a state transition xy, ρ is the pheromone evaporation coefficient and $\Delta\tau_{xy}^k$ is the amount of pheromone deposited by k^{th} ant,

$$\Delta\tau_{xy}^k = \begin{cases} Q/L_k, & \text{if ant k uses curve xy in its tour} \\ 0, & \text{orelse} \end{cases} \qquad (4.10)$$

In the above equation 4.10, L_k is the cost of the kth ant's tour (length) and Q is the constant. The flow of this work is explained in the following Figure 4.1.

ACO algorithm for task scheduling

Input: clustered employees, task list,

Output: Scheduled tasks

Step 1: Initialization:

    Time t=0, number of cycles NC =0, pheromone $\tau_{xy}(t) = c$, positioning of m ants

Step 2: Initialize tabu list which is a dynamically growing vector

Step 3: Each ant iteratively schedule the tasks to the clustered employees

Step 4: Calculate the fitness value of each ant (which is briefly explained in section 3.4) by using equation 3.10

Step 5: Update each ant based on the fitness value

Step 6: Calculate $\Delta\tau_{xy}^k$ using equation 4.10

Step 7: Update $\tau_{xy}(t + n)$

Step 8: Increment discrete time as t=t+n and NC-NC+1

Step 9: If (NC<NC_{max}) then go to step 2

Step 10: Stop the process

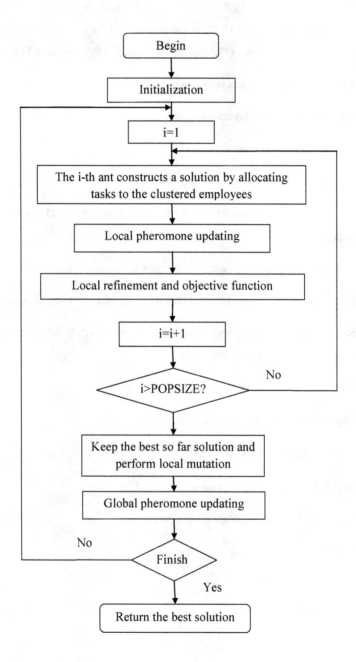

Figure 4.1   Flowchart of ACO Based task scheduling method

4.3 RESULT AND DISCUSSION

In this section, the experimental results of the proposed approach are illustrated.

4.3.1 Project Completion Time versus Project Completion Cost

The objective function is minimizing the project completion cost and completion time of the software project. Hence, the performance of the proposed approach is analyzed with other methods in terms of time and cost. The following Table 4.1 shows the comparison of Project Completion Time versus Project Completion Cost for the PSPLIB dataset mentioned in chapter 3.3.1.

Table 4.1 Comparison of Project Completion Time versus Project Completion Cost

Project Completion Time (ms)	Project Completion Cost (Rs)	
	ACO-UH	FCM-ACO-UH
500	5350	5300
1000	5425	5398
1500	5550	5460
2000	5692	5582

2500	5750	5636
3000	5850	5700
3500	5950	5786
4000	6050	5829
4500	6150	5860

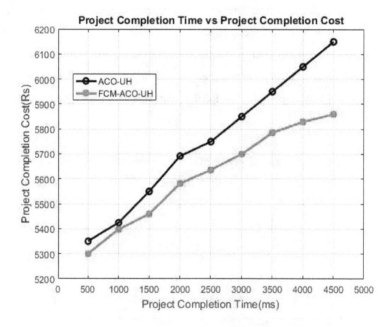

Figure 4.2 Comparison of Project Completion Time versus Project Completion Cost

Figure 4.2 shows that the comparison results of the proposed and existing approaches in terms of cost and time parameters. The time taken to schedule and resource allocation in the project is very low and the cost taken for scheduling and resource allocation is also very low in the proposed system.

4.3.2 Project Completion Time

The project completion time is defined as the amount of time taken to complete tasks successfully in a project. The following table 4.2 shows the comparison of project completion time.

Table 4.2   Comparison of Project Completion Time

	Methods	
	ACO-UH	FCM-ACO-UH
Project Completion Time (ms)	6100	5647

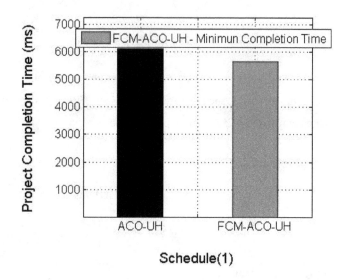

Figure 4.3   Comparison of Project Completion Time

Figure 4.3 shows that the comparison results of the proposed and existing approaches in terms of project completion time. The time taken to schedule and resource allocation in the project is very low in the proposed system.

4.3.3   Project Completion Cost

The project completion cost is the total amount cost required for a project. The following table 4.3 shows the comparison of project completion cost.

Table 4.3   Comparison of Project Completion cost

	Methods

	ACO-UH	FCM-ACO-UH
Project Completion Cost (Rs)	6000	5680

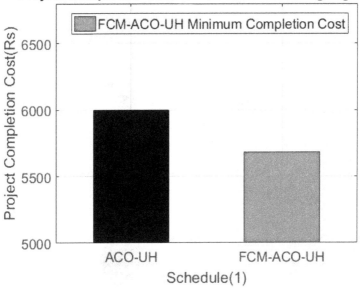

Figure 4.4   Comparison of Project Completion cost

Figure 4.4 shows that the comparison results of the proposed and existing approaches in terms of project completion cost. The cost to schedule and resource allocation in the project is very low taken for scheduling and resource allocation is very low in the proposed system.

4.3.4 Project Allocation Time

The project allocation time is defined as the amount of time taken by each method to schedule tasks of software projects. The following table 4.4 shows the comparison of project allocation time.

Table 4.4   Comparison of Project Allocation Time

	Methods	
	ACO-UH	FCM-ACO-UH
Project Allocation Time (ms)	5421	5014

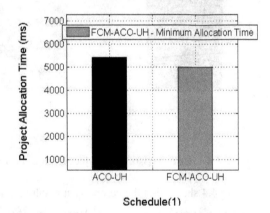

Figure 4.5   Comparison of Project Allocation Time

Figure 4.5 shows that the comparison results of the proposed and existing approaches in terms of project allocation time. The time taken to allocate the tasks to the appropriate employees by the proposed method is very low in the proposed system.

4.3.5 Productivity

Software productivity is typically measured by the project size divided by effort. Effort is commonly expressed in man-months of labour. It can be mathematically expressed as follows:

$$Productivity = \frac{Number\ of\ function\ points}{Effort\ in\ man\ months}$$

The following Table 4.5 shows the comparison of productivity.

Table 4.5   Comparison of Productivity

	Methods	
	ACO-UH	FCM-ACO-UH
Productivity	90	94

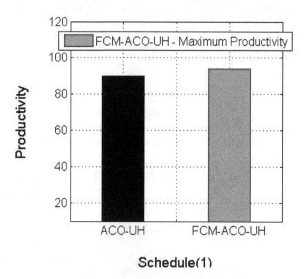

Figure 4.6　Comparison of Productivity

Figure 4.6 shows that the comparison results of the proposed and existing approaches in terms of productivity. The productivity of project scheduling and resource allocation is very high for the proposed approach.

4.4　SUMMARY

In this chapter, the problem of flexibility of the search space and human resources in project scheduling are considered. These problems are resolved by the proposed Fuzzy C Means clustering with Ant Colony Optimization method. Initially the proposed method introduced FCM clustering method for event based scheduler. In FCM, the similar features and experienced employee information are grouped into a cluster, that improved the results of Event Based Scheduler and then execute the ACO method to solve the scheduling problem. Thus, the proposed method reduces the search space and increased resource flexibility of the project scheduling process. The experimental results show that the proposed FCM-ACO-UH improves the productivity such as 4.49% than ACO-UH as well as reduces the project allocation time such as 7.52% than ACO-UH. In addition, this approach reduces the project completion cost such as 2.35% than ACO-UH technique. Hence, the results prove that the proposed FCM-ACO-UH has a better performance in terms of minimum cost and time compared to the other software project scheduling techniques.

CHAPTER 5

FUZZY C MEANS CLUSTERING WITH ABC FOR FAST AND EFFICIENT SCHEDULING AND RESOURCE ALLOCATION

This chapter provides detailed information about the Fuzzy C means with Artificial Bee Colony method for fast and efficient software project scheduling and resource allocation. This chapter describes how the proposed Fuzzy C means with ABC reduce the search operations and improve search retrieval of data.

5.1 INTRODUCTION

Successful Management of complex software relies on the process to solve complex problems involved in the software management process. In the FCM with ACO based task scheduling technique considered the user acquaintances while performing the event based scheduling task. In addition to that, it considered both the employee experience and other expertise for scheduling tasks. The FCM method clustered similar employees based on the working skills and their experience. Then the tasks in the task list were assigned to the clustered employees based on the ACO method. In order to reduce the search operations and improve the retrieval of search data, Artificial Bee Colony (ABC) is proposed in this research work. In this proposed work the clustered employees are given as input to the ABC algorithm which scheduled the tasks to the appropriate employees. The ABC algorithm (Karaboga, D., & Basturk, B. 2007) is to stimulate the foraging behavior of bees. In this proposed work, ABC has been efficaciously utilized for project scheduling problem, as it is easy to develop and fix many optimization problems with only a few controls of parameters. Thus, the proposed FCM with ABC based task scheduling process solved the scheduling problem with less search operation process. The experimental results shows the comparison of FCM with ACO and FCM with ABC based task scheduling process in terms of cost versus time and throughput which prove the effectiveness of the proposed FCM with ABC based task scheduling process.

5.2 FCM WITH ABC BASED TASK SCHEDULING

FCM clustering methods for event based task scheduler clustered similar employees based on similar features and experienced employee information. Later, the method accomplishes the Artificial Bee Colony optimization method to solve the intricate planning problem. In ABC, the colony of artificial bees consists of three groups of bees: employee bee, onlooker bee and scout. The employee bee visits the food source position based on the information of task list allocation to employee and the similar employee cluster experience from the FCM clustering results. Each one of the employee bee in ABC fold the number of tasks owed to user and users' experience list. The bees are accommodated according to the performance, based on the project planning data. These bees execute the local investigation for estimation of shortest possible makespan of each task with the best of the experience level to each employee in the clustered group. Moreover these bees exploit the nearest neighboring location results of the scheduling food source. The software project problem was unraveled by onlooker bees which waited in the nest area. The solution for software project planning is made, based on the experience of employee to work for the specified task which is given by employee bee. The global investigation is processed by the onlooker bees which solved the software project planning and improved global optimum software project scheduling results. The Scout bee in ABC is utilized to discover the new employee task list and planned experienced employee allocation matrix which is not focused on by the employee bee. This employee bee, onlooker bee and scout bee are processed iteratively until there is an achievement of a maximum number of iterations termination criterions. The fitness (objective function) value for each employee task list for software project scheduling problem was computed based on the following parameters.

$\gamma_e^{(i,j)}$ denotes the heuristic of choosing the i-th employee to work for the task t_j. $\gamma_e^{(i,j)}$ denotes the rate between the proficiency p_{ij} of i-th employee for the task t_j and the hour salary S_i of the employee. $\gamma_e^{(i,j)}$ is denoted as follows:

$$\gamma_e^{(i,j)} = \frac{p_{ij}}{S_i} \qquad (5.1)$$

The higher proficiency employee with lower salary is more probable to be chosen to work for t_j. If the experience of an employee on a task $join_i$ at the beginning of a time unit and they work for a fraction of the time (b) on this task:

$$\delta_e(i,j) = \frac{Taskexp_e}{S_i \times Lthr} \qquad (5.2)$$

In the above equation 5.2, Lthr denotes the learning speed threshold value. These parameters are those belong to cluster employee attributes. Then the employee allocation matrix is constructed based on the following steps:

Initially set all values in the employee allocation matrix to 0

For each task $t_j (j = 1,2, ..., m)$ describe the workloads for t_j by the following sub steps:

Compute the value of $\gamma_e^{(i,j)}$ and $\delta_e(i,j)$ for all employees

Check on fitness value for each employee bees by using following equation 5.3:

$$f_i = \frac{1}{1+f_i} \qquad (5.3)$$

The fitness of each employee bee is calculated based on the parameters in equation 5.2. In the following equation 5.4, an artificial onlooker bee estimated probability value p_i for software project planning is given as follows:

$$p_i = \frac{f_i}{\sum_{n=1}^{N} f_n} \qquad (5.4)$$

In the above equation 5.4, f_i signifies the fitness value of the task to each employee i in the location and N denotes the size of the population. The nominated task position is updated by using the following equation 6.5

$$v_{ij} = x_{ij} + \theta_{ij}(x_{ij} - x_{kj}) \qquad (5.5)$$

In the above equation 5.5, k and j variables are randomly selected for the different tasks $\varepsilon \{1,2,...N\}$ and j $\varepsilon \{1,2,...D\}$. $\Phi_{ij}\varepsilon[-1,1]$. If the value of x_{ij} exceeds its threshold value, the scheduling problem is acceptable, otherwise it is not acceptable as best scheduling results. It is also replaced by scout bees. As per the procedure of ABC, if position of a bee does not produce best results within a pre specified number of iterations, then the current task position is considered to be neglected and it is updated by using the following equation:

$$x_i^j = x_{min}^j + rand(0,1)(x_{max}^j - x_{min}^j) \qquad (5.6)$$

The above defined equation depends mostly on following the parameters that restrict the operation N, Maximum Number of Cycles (MNC).

The various ABC variants were developed and the parameters assigned in the starting stage and in building the plans according to the problems by ants and have been processed on further stages. The project planning model planned the number of stages for the process based on the model proposed and chosen to obtain the best result.

The following algorithm shows the optimization of ABC for software project scheduling. Initially they initialized and appraised the population of ABC. Then, the cycle value is assigned as 1and repeat the process until the initialization is done. The new software project scheduling is produced and the greedy selection process is evaluated. Finally, computing the probability values, producing the software project scheduling solutions, applying the greedy selection process and abandoning the task results in the scout and memorizing the best solution using cycle=cycle+1 will be done iteratively until the cycle value is reached Maximum Cycle Number (MCN).

Artificial Bee Colony optimization algorithm for software project scheduling

Input: Clustered employees, employee attributes, task attributes

Output: Scheduled tasks

Step 1: Initializepopulation of solutions $x_i, i = 1,2, ..., N$, each population as number of tasks for each employee

Step 2: Appraise the population

Step 3: Set cycle =1

Step 4: Repeat

Step 5: Generate new software project scheduling v_i for employee bees (tasks) by using equation 5.4

Step 6: Process the greedy selection for employee bees

Step 7: Evaluate the probability values p_i for software project scheduling solutions x_i by using equation 5.5

Step 8: Produce the software project scheduling solutions v_i for the onlookers from the solutions x_i designated depending on p_i and compute them

Step 9: Apply the greedy selection process for the onlookers

Step 10: Adopt the abandoned task result in the scout, if it exists and replace it with a new randomly produced solution by using the equation 5.5

Step 11: Memorize the best solution achieved so far

Step 12: cycle=cycle+1

Step 13: until cycle=MCN

5.3 RESULT AND DISCUSSION

5.3.1 Project Completion Time versus Project Completion Cost

The objective function is minimizing the project completion cost and completion time of the software project. Hence, the performance of the proposed approach is analyzed with other methods in terms of cost and time. The following Table 5.1 shows the comparison of Project Completion Time versus Project Completion Cost for the PSPLIB dataset mentioned in chapter 3.3.1.

Table 5.1   Comparison of Project Completion Time versus Project Completion Cost

Project Completion Time (ms)	Project Completion Cost (Rs)	
	FCM-ACO-UH	FCM-ABC-UH
500	5300	5250
1000	5398	5340
1500	5460	5410
2000	5582	5490
2500	5636	5550
3000	5700	5610
3500	5786	5700
4000	5829	5780
4500	5860	5800

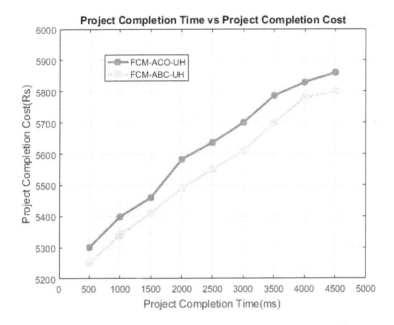

Figure 5.1   Comparison of Project Completion Time versus Project Completion Cost

Figure 5.1 shows that the comparison results of the proposed approach and existing approach in terms of Project Completion Time and Project Completion Cost parameters. The time taken to schedule and resource allocation of the project is very low and the cost taken for scheduling and resource allocation is also very low in the proposed system.

5.3.2   Project Completion Time

The project completion time is defined as the amount of time taken to complete tasks successfully in a project. The following table 5.2 shows the comparison of project completion time.

Table 5.2   Comparison of Project Completion Time

	Methods	
	FCM-ACO-UH	FCM-ABC-UH
Project Completion Time (ms)	5647	5104

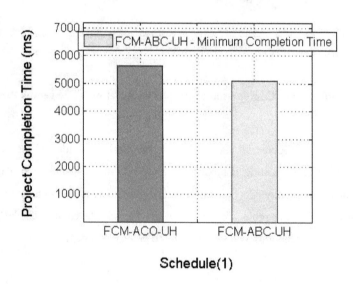

Figure 5.2   Comparison of Project Completion Time

Figure 5.2 shows the comparison results of the proposed approach and the existing approach in terms of project completion time. The time taken to schedule and resource allocation in the project is very low in the proposed system.

5.3.3 Project Completion Cost

The project completion cost is the total amount required for a project. The following table 5.3 shows the comparison of project completion cost.

Table 5.3   Comparison of Project Completion cost

	Methods	
	FCM-ACO-UH	FCM-ABC-UH
Project Completion Cost (Rs)	5680	5600

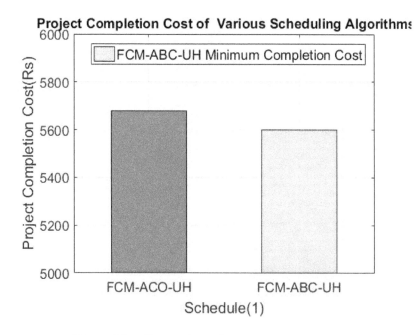

Figure 5.3   Comparison of Project Completion cost

Figure 5.3 shows that the comparison results of the proposed approach and the existing approach in terms of project completion cost. The cost to schedule and resource allocation in the project is very low and the cost taken for scheduling and resource allocation is very low in the proposed system.

5.3.4 Project Allocation Time

The project allocation time is defined as the amount of time taken by each method to schedule various tasks in software projects. The following table 5.4 shows the comparison of project allocation time.

Table 5.4   Comparison of Project Allocation Time

	Methods	
	FCM-ACO-UH	FCM-ABC-UH
Project Allocation Time (ms)	5014	4517

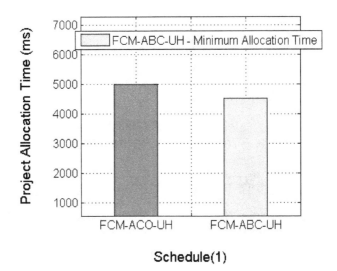

Figure 5.4   Comparison of Project Allocation Time

Figure 5.4 shows that the comparison results of proposed approach and existing approach in terms of project allocation time. The time taken to allocate the tasks to the appropriate employees by the proposed method is very low in the proposed system.

5.3.5   Productivity

Software productivity is typically measured by the project size divided by effort. Effort is commonly expressed in man-months of labor. It can be mathematically expressed as follows:

$$Productivity = \frac{Number\ of\ function\ points}{Effort\ in\ man\ months}$$

The following Table 5.5 shows the comparison of productivity.

Table 5.5   Comparison of Productivity

	Methods	
	FCM-ACO-UH	FCM-ABC-UH
Productivity	94	97

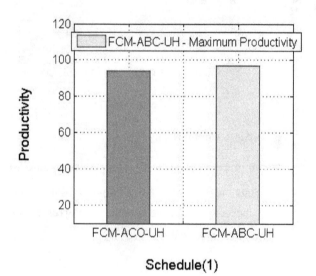

Figure 5.5   Comparison of Productivity

Figure 5.5 shows the comparison results of proposed approach and existing approach in terms of productivity. The productivity of project scheduling and resource allocation is very high for the proposed approach.

5.4 OVERALL RESULT AND DISCUSSION

In this section, the experimental results are conducted between existing software task scheduling methods such as TS, KGA and proposed methods are ACO-UH, FCM-ACO-UH, FCM-ABC-UH in terms of project completion time vs. project completion cost, project completion time, project completion cost, project allocation time and productivity.

5.4.1 Project Completion Time Vs. Project Completion Cost

The objective function is minimizing the project completion time and completion cost of the software project. Hence, the performance of the proposed approach is analyzed with other methods in terms of cost and time. The following Table 5.6 shows the comparison of Project Completion Time versus Project Completion Cost.

Table 5.6   Comparison of Project Completion Time versus Project Completion Cost

Project Completion Time (ms)	Project Completion Cost (Rs)					
	TS	KGA	ACO-L	ACO-UH	FCM-ACO-UH	FCM-ABC-UH
500	6100	5800	5400	5350	5300	5250
1000	6220	5980	5510	5425	5398	5340
1500	6370	6060	5620	5550	5460	5410
2000	6450	6280	5860	5692	5582	5490

2500	6600	6390	5940	5750	5636	5550
3000	6720	6500	6020	5850	5700	5610
3500	6800	6600	6100	5950	5786	5700
4000	6840	6710	6180	6050	5829	5780
4500	6900	6800	6240	6150	5860	5800

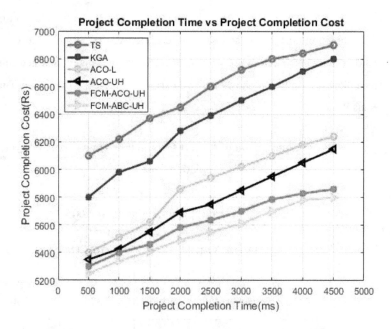

Figure 5.6 Comparison of Project Completion Time versus Project Completion Cost

Figure 5.6 shows that the comparison results of the proposed approach and the existing approach in terms of cost and time parameters. The time taken to schedule and resource allocation in the project is very low and the cost taken for scheduling and resource allocation is also very low in the FCM-ABC-UH proposed system. In the graph, the project completion time in terms of milliseconds is taken in x-axis and the project completion cost in terms of $ is taken in y-axis. When the project completion time is 4500, the project completion cost of FCM-ABC-UH is 1.02% lower than FCM-ACO-UH, 5.69% lower than ACO-UH, 7.05% lower than ACO-L, 14.71% lower than KGA and 15.94% lower than TS approach. It shows that the proposed FCM-ABC-UH approach has better completion cost for different project completion time.

5.4.2 Project Completion Time

The project completion time is defined as the amount of time taken to complete tasks successfully in a project. The following Table 5.7 shows the comparison of project completion time.

Table 5.7   Comparison of Project Completion Time

	Methods					
	TS	KGA	ACO-L	ACO-UH	FCM-ACO-UH	FCM-ABC-UH
Project Completion Time (ms)	7000	6800	6500	6100	5647	5104

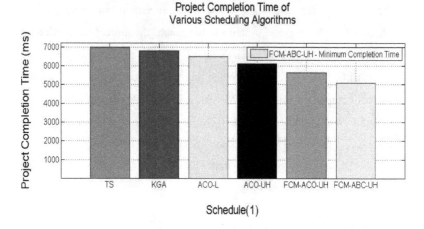

Figure 5.7   Comparison of Project Completion Time

Figure 5.7 shows that the comparison results of the proposed and existing approaches in terms of project completion time. The time taken to schedule and resource allocation in the project is very low in the proposed system. The project completion time of proposed FCM-ABC-UH approach is 9.62% decreased than FCM-ACO-UH, 16.32% decreased than ACO-UH, 21.48% decreased than ACO-L, 24.94% decreased than KGA and 27.09% decreased than TS approach. From the above values, it is noted that the proposed FCM-ABC-UH approach outperforms than the other approaches.

5.4.3   Project Completion Cost

The project completion cost is the total amount of cost required for a project. The following table 5.8 shows the comparison of project completion cost.

Table 5.8   Comparison of Project Completion cost

	Methods					
	TS	KGA	ACO-L	ACO-UH	FCM-ACO-UH	FCM-ABC-UH
Project Completion Cost (Rs)	6650	6400	6140	6000	5680	5600

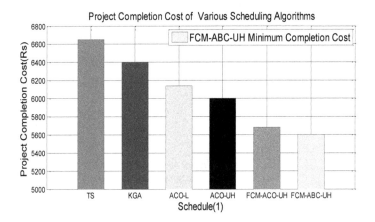

Figure 5.8   Comparison of Project Completion cost

Figure 5.8 shows that the comparison results of the proposed and existing approaches in terms of project completion cost. The cost to schedule and resource allocation in the project taken is very low for scheduling and resource allocation is very low in the proposed system. The project completion cost of proposed FCM-ABC-UH approach is 1.41% lower than FCM-ACO-UH, 6.67% lower than ACO-UH, 8.79% lower than ACO-L, 12.5% lower than KGA and 15.79% lower than TS approach. As a result, it is observed that the proposed FCM-ABC-UH approach has better project completion cost than the other approaches.

5.4.4 Project Allocation Time

The project allocation time is defined as the amount of time taken by each method to schedule tasks of software projects. The following Table 5.9 shows the comparison of project allocation time.

Table 5.9 Comparison of Project Allocation Time

	Methods					
	TS	KGA	ACO-L	ACO-UH	FCM-ACO-UH	FCM-ABC-UH
Project Allocation Time (ms)	6870	6500	6147	5421	5014	4517

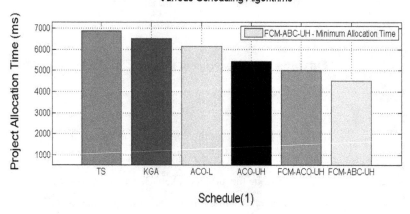

Figure 5.9   Comparison of Project Allocation Time

Figure 5.9 shows that the comparison results of the proposed approach and the existing approach in terms of project allocation time. The time taken to allocate the tasks to the appropriate employees by the proposed method is very low in the proposed system. The project allocation time of proposed FCM-ABC-UH is 9.91% lower than FCM-ACO-UH, 16.68% lower than ACO-UH, 26.52% lower than ACO-L, 30.51% lower than KGA and 34.25% lower than TS approach. Based on this observation, it is identified that the proposed FCM-ABC-UH approach has better project allocation time than the other approaches.

5.4.5 Productivity

Software productivity is typically measured by project size divided by effort. Effort is commonly expressed in man-months of labor. It can be mathematically expressed as follows:

$$Productivity = \frac{Number\ of\ function\ points}{Effort\ in\ man\ months}$$

Table 5.10   Comparison of Productivity

	Methods					
	TS	KGA	ACO-L	ACO-UH	FCM-ACO-UH	FCM-ABC-UH
Productivity	80	85	88	90	94	97

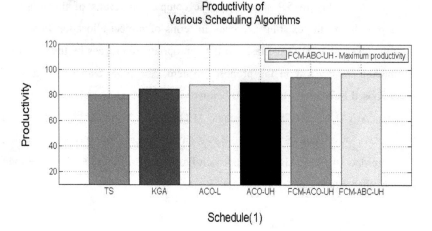

Figure 5.10 Comparison of Productivity

Figure 5.10 shows that the comparison results of the proposed approach and the existing approach in terms of productivity. The productivity of project scheduling and resource allocation is very high for the FCM-ABC-UH proposed approach. The productivity of proposed FCM-ABC-UH approach is 3.09% greater than FCM-ACO-UH, 7.22% greater than ACO-UH, 9.28% greater than ACO-L, 12.37% greater than KGA and 17.53% greater than TS approach. By using this observation, it is known that the proposed FCM-ABC-UH approach has better productivity than the other approaches.

5.5　SUMMARY

In this chapter, the problem of search operation in software project scheduling is considered. This problem is solved by proposed Fuzzy C Means clustering with Artificial Bee Colony method. Initially, similar featured employees are clustered using FCM method. Then the tasks of the software projects are scheduled based on the proposed ABC method. In this method, the employee bee, onlooker bee and scout bee is employed in ABC method to effectively schedule the tasks to the more appropriate employees in the clusters. The proposed method influences the better plans with lower costs in short time and produce unwavering employee appointments, collated with distinct subsistent tactics, since it considers the characteristics of experience additionally. The experimental results are conducted to prove the effectiveness of the proposed work in the terms of cost versus time and productivity. The experimental results show that the proposed FCM-ABC-UH improves the productivity such as 3.19% than FCM-ACO-UH , as well as reduces the project allocation time such as 9.91% than FCM-ACO-UH. In addition, this approach reduces the project completion cost such as 1.23% than FCM-ACO-UH technique. Hence, the results prove that the proposed FCM-ABC-UH has a better performance in terms of minimum cost and time compared to the other software project scheduling techniques. Moreover, in this chapter the various project scheduling and resource allocation approaches are analyzed in terms of Project Completion Time versus Project Completion Cost, Project completion time, Project completion cost, Project allocation time and productivity in order to prove the effectiveness of the proposed project scheduling and resource allocation method. TS, KGA and ACO-L are compared with the proposed methods ACO-UH, FCM-ACO-UH, FCM-ABC-UH. From the analysis, it is proved that the proposed FCM-ABC-UH approach has better Project Completion Time versus Project Completion Cost, Project completion time, Project completion cost, Project allocation time and productivity than the other approaches.

CHAPTER 6

CONCLUSION AND FUTURE WORK

6.1 CONCLUSION

Nowadays, software companies are facing a distinctly competitive marketplace due to the fast improvement of the software industry. The companies have to provide proficient project plans to minimize the software construction cost. However, the problem of project planning and scheduling are the most complex and challenging in large-scale projects. The project manager requires estimating the project workload and cost and deciding the project schedule and resource allocation for planning the software project. Hence, in this research, efficient heuristic computational strategies are proposed for achieving effective project scheduling and resource allocation.

In the first work of the research, a novel method is proposed which resolves the software project planning issues. Initially, an event-based scheduler and ACO algorithm are introduced which is used to develop the flexible and effective model. The proposed algorithm is used for representing the project scheduling based on the task list and a planned employee allocation matrix. The proposed event-based scheduler is utilized for adjusting the allocation of employees during certain as well as uncertain events. Thus, the proposed approach helps software development managers manage with the uncertainties they face during resource allocation and project scheduling. Furthermore, a decision tool is developed for helping the managers by replicating the various uncertainties that will unfold and the means to interact as the project progresses.

In the second work of the research work, the resource flexibility problem in software project scheduling was resolved by proposed Fuzzy C Means clustering with ACO method. By appointing the employee to only one task, the flexibility of human resources can be decreased which is considered in this work. Moreover, the user acquaintances were not considered in the previous work while performing the event based scheduling tasks. The proposed FCM based EBS considers the employee experience and other expertise for scheduling task by proposed paradigm FCM clustering method. The FCM method, group the analogous user experience based on the working skills and experience. Then the project planning problem in the EBS is solved by using ACO algorithm. Thus, the proposed method, effectively scheduled the tasks of projects to the more appropriate employees by solving search space problem and increasing the resource flexibility.

In the third work of the research work, the search space problem is resolved by proposed FCM with Artificial Bee Colony (ABC) method. The similar featured employees are clustered using FCM clustering. Then the ABC algorithm is used to reduce the search space of the software project scheduling process. In ABC method, the employed the employee bee, onlooker bee and scout bee to decipher the software project planning problem. Tentative results prove that the representation, in the anticipated work, the similar features and experienced employee's information are convened into the same cluster to improve the results of event based scheduler then execute the ABC method to solve the scheduling problem. Thus, the proposed method influences better plans with lower costs in a short time.

6.2 FUTURE WORK

The future extension of this research may focus on an efficient optimal task scheduling could be enhanced by novel optimization methods such as Cat Swarm Optimization, Cockroach Swarm Optimization, Bean Optimization, etc., which may provide the best and fastest results. Better system dynamics would be combined with the modelling of progress of tasks where better experience models and training models will be encompassed. Furthermore, the impact of team size will be investigated by these topics.

There is also a need for further examination is concerned with the better know-how of developer productivity so that it will help extra accurate estimations of the period and cost of obligations. Since the subject of productivity might be very substantial, an intensive research will be required to make certain that its utilization inside the optimization strategies contributes to the technology of practical solutions. Additional constraints can be included in the heuristic function to cater to any other practical difficulties faced by the project managers.

The basic software structure is able of handling multiple resources and research will be required regarding how best to manage multiple resources simultaneously. It would also be of interest to consider how teamwork deviation information may be fed back to the earlier stages of project selection and scheduling decision.

It would be desirable to develop special algorithms to compute the optimal solution to the variable intensity and the other with fixed intensity to the preemptive tasks which also fit into real-world scenarios and also need a model for analyzing the difference between the planned and actual resource bookings and find out the reasons for variations that are helpful to improve the process further.

Future work could, therefore, focus on incorporating degrees of resource dedication and availability, as well as resource leveling constraints for multi-project task scheduling. In addition, it would also be interesting to research the allocation of resources and scheduling of tasks for distributed software development projects.

Tracking Employee Training:

The creation and implementation of training programs are important in enhancing the human resources knowledge. To ensure that staffs is fully qualified for their jobs, The Employee training programs are implemented by company which is used for job advancement and employees are required to complete certain certification programs in some job settings. Tracking Employee Training system is important to track the failure of employees in their qualification and certification.

More constraints:

Future work could consider such additional objectives. For example, we could consider capability constraints and availability constraints, in scheduling. Other constraints, such as different activities needing the same resource, will be modeled in future work. Find out the general conditions under which prioritization of projects are considered that would lead to optimal solution.

Programmer Productivity:

Future enhancements concerns empirical dynamic models of the software process to address additional communication factors, such as the programmer productivity.

Computing for Deadlines:

It would be useful to compute deadlines of the project that meet the requirement of minimized makespan with importance of each objective specified by the user.

There is also a want to explore the rates of completion of software development tasks at specific levels of productiveness, and to and to check whether or not or not a saturation issue exits in which a mission cannot be finished faster irrespective of the charge of productivity.

REFERENCES

1. Alves, LM, Oliveira, S, Ribeiro, P & MacHado, RJ 2014, 'An empirical study on the estimation of size and complexity of software applications with function points analysis', Proceedings - 14th International Conference on Computational Science and Its Applications, ICCSA 2014, pp. 27-34.

2. Amiri, M & Barbin, JP 2015, 'New Approach for Solving Software Project Scheduling Problem Using Differential Evolution Algorithm', International Journal in Foundations of Computer Science & Technology, vol. 5, no. 1, pp. 1-9.

3. Amirian, H & Sahraeian, R 2017, 'Solving a grey project selection scheduling using a simulated shuffled frog leaping algorithm', Computers and Industrial Engineering, vol. 107, pp. 141-149.

4. Anwar, Z, Bibi, N & Ahsan, A 2013, 'Expertise based skill management model for effective project resource allocation under stress in software industry of Pakistan', Proceedings of 2013 6th International Conference on Information Management, Innovation Management and Industrial Engineering, ICIII 2013, vol. 1, pp. 509-513.

5. Avinash, A & Ramani, K 2014, 'A Hybrid Technique for Software Project Scheduling and Human Resource Allocation', International Journal of Engineering Development and Research, vol. 2, no. 3, pp. 3243-3251.

6. Azzeh, MYa, Neagu, D & Cowling, PI 2008, 'Software Project Similarity Measurement Based on Fuzzy C-Means', Making Globally Distributed Software Development A Success Story, Lecture Notes in Computer Science, LNCS5007, pp. 123-134.

7. Bhele, NR, Dhamale, SS & Parkhe, B 2015, 'A Proposed Scheme for Software Project Scheduling and Allocation with Event Based Scheduler using Ant Colony Optimization', vol. 4, no. 1, pp. 147-151.

8. Bibi, N, Ahsan, A & Anwar, Z 2014, 'Project Resource Allocation Optimization using Search Based Software Engineering – A Framework', International Conference in Digital Information Management (ICDIM), pp. 226-229.

CHAPTER 6

CONCLUSION AND FUTURE WORK

6.1 CONCLUSION

Nowadays, software companies are facing a distinctly competitive marketplace due to the fast improvement of the software industry. The companies have to provide proficient project plans to minimize the software construction cost. However, the problem of project planning and scheduling are the most complex and challenging in large-scale projects. The project manager requires estimating the project workload and cost and deciding the project schedule and resource allocation for planning the software project. Hence, in this research, efficient heuristic computational strategies are proposed for achieving effective project scheduling and resource allocation.

In the first work of the research, a novel method is proposed which resolves the software project planning issues. Initially, an event-based scheduler and ACO algorithm are introduced which is used to develop the flexible and effective model. The proposed algorithm is used for representing the project scheduling based on the task list and a planned employee allocation matrix. The proposed event-based scheduler is utilized for adjusting the allocation of employees during certain as well as uncertain events. Thus, the proposed approach helps software development managers manage with the uncertainties they face during resource allocation and project scheduling. Furthermore, a decision tool is developed for helping the managers by replicating the various uncertainties that will unfold and the means to interact as the project progresses.

In the second work of the research work, the resource flexibility problem in software project scheduling was resolved by proposed Fuzzy C Means clustering with ACO method. By appointing the employee to only one task, the flexibility of human resources can be decreased which is considered in this work. Moreover, the user acquaintances were not considered in the previous work while performing the event based scheduling tasks. The proposed FCM based EBS considers the employee experience and other expertise for scheduling task by proposed paradigm FCM clustering method. The FCM method, group the analogous user experience based on the working skills and experience. Then the project planning problem in the EBS is solved by using ACO algorithm. Thus, the proposed method, effectively scheduled the tasks of projects to the more appropriate employees by solving search space problem and increasing the resource flexibility.

In the third work of the research work, the search space problem is resolved by proposed FCM with Artificial Bee Colony (ABC) method. The similar featured employees are clustered using FCM clustering. Then the ABC algorithm is used to reduce the search space of the software project scheduling process. In ABC method, the employed the employee bee, onlooker bee and scout bee to decipher the software project planning problem. Tentative results prove that the representation, in the anticipated work, the similar features and experienced employee's information are convened into the same cluster to improve the results of event based scheduler then execute the ABC method to solve the scheduling problem. Thus, the proposed method influences better plans with lower costs in a short time.

6.2 FUTURE WORK

The future extension of this research may focus on an efficient optimal task scheduling could be enhanced by novel optimization methods such as Cat Swarm Optimization, Cockroach Swarm Optimization, Bean Optimization, etc., which may provide the best and fastest results. Better system dynamics would be combined with the modelling of progress of tasks where better experience models and training models will be encompassed. Furthermore, the impact of team size will be investigated by these topics.

There is also a need for further examination is concerned with the better know-how of developer productivity so that it will help extra accurate estimations of the period and cost of obligations. Since the subject of productivity might be very substantial, an intensive research will be required to make certain that its utilization inside the optimization strategies contributes to the technology of practical solutions. Additional constraints can be included in the heuristic function to cater to any other practical difficulties faced by the project managers.

The basic software structure is able of handling multiple resources and research will be required regarding how best to manage multiple resources simultaneously. It would also be of interest to consider how teamwork deviation information may be fed back to the earlier stages of project selection and scheduling decision.

It would be desirable to develop special algorithms to compute the optimal solution to the variable intensity and the other with fixed intensity to the preemptive tasks which also fit into real-world scenarios and also need a model for analyzing the difference between the planned and actual resource bookings and find out the reasons for variations that are helpful to improve the process further.

Future work could, therefore, focus on incorporating degrees of resource dedication and availability, as well as resource leveling constraints for multi-project task scheduling. In addition, it would also be interesting to research the allocation of resources and scheduling of tasks for distributed software development projects.

Tracking Employee Training:

The creation and implementation of training programs are important in enhancing the human resources knowledge. To ensure that staffs is fully qualified for their jobs, The Employee training programs are implemented by company which is used for job advancement and employees are required to complete certain certification programs in some job settings. Tracking Employee Training system is important to track the failure of employees in their qualification and certification.

More constraints:

Future work could consider such additional objectives. For example, we could consider capability constraints and availability constraints, in scheduling. Other constraints, such as different activities needing the same resource, will be modeled in future work. Find out the general conditions under which prioritization of projects are considered that would lead to optimal solution.

Programmer Productivity:

Future enhancements concerns empirical dynamic models of the software process to address additional communication factors, such as the programmer productivity.

Computing for Deadlines:

It would be useful to compute deadlines of the project that meet the requirement of minimized makespan with importance of each objective specified by the user.

There is also a want to explore the rates of completion of software development tasks at specific levels of productiveness, and to and to check whether or not or not a saturation issue exits in which a mission cannot be finished faster irrespective of the charge of productivity.

Printed in the USA
CPSIA information can be obtained
at www.ICGtesting.com
LVHW011738230923
759042LV00003B/624